The Metabolic Role of Peroxisome in Health and Disease

Edited by Hasan Basri İla

Published in London, United Kingdom

The Metabolic Role of Peroxisome in Health and Disease
http://dx.doi.org/10.5772/intechopen.95145
Edited by Hasan Basri İla

Contributors
Naomi van Wijk, Michal Linial, Hasan Basri İla, Meimei Wang, Yakun Liu, Ni Chen, Ye Zhao, Juan Wang, Uma V. Manjappara, Beekanahalli G. Mallikarjuna, Shazia Usmani, Abdul Hafeez, Shadma Wahab, Shabana Khatoon, Syed Misbahul Hasan, Nishat Fatima

First published in London, United Kingdom, 2022 by IntechOpen
IntechOpen is the global imprint of INTECHOPEN LIMITED, registered in England and Wales, registration number: 11086078, 5 Princes Gate Court, London, SW7 2QJ, United Kingdom

British Library Cataloguing-in-Publication Data
A catalogue record for this book is available from the British Library

Additional hard and PDF copies can be obtained from orders@intechopen.com

The Metabolic Role of Peroxisome in Health and Disease
Edited by Hasan Basri İla
p. cm.

This title is part of the Physiology Book Series, Volume 17
Topic: Cell Physiology
Series Editor: Tomasz Brzozowski
Topic Editor: Angel Catala

Print ISBN 978-1-80355-918-6
Online ISBN 978-1-80355-919-3
eBook (PDF) ISBN 978-1-80355-920-9
ISSN 2631-8261

IntechOpen Book Series
Physiology
Volume 17

Aims and Scope of the Series

Modern physiology requires a comprehensive understanding of the integration of tissues and organs throughout the mammalian body, including the cooperation between structure and function at the cellular and molecular levels governed by gene and protein expression. While a daunting task, learning is facilitated by identifying common and effective signaling pathways mediated by a variety of factors employed by nature to preserve and sustain homeostatic life. As a leading example, the cellular interaction between intracellular concentration of Ca+2 increases, and changes in plasma membrane potential is integral for coordinating blood flow, governing the exocytosis of neurotransmitters, and modulating gene expression and cell effector secretory functions. Furthermore, in this manner, understanding the systemic interaction between the cardiovascular and nervous systems has become more important than ever as human populations' life prolongation, aging and mechanisms of cellular oxidative signaling are utilised for sustaining life. Altogether, physiological research enables our identification of distinct and precise points of transition from health to the development of multimorbidity throughout the inevitable aging disorders (e.g., diabetes, hypertension, chronic kidney disease, heart failure, peptic ulcer, inflammatory bowel disease, age-related macular degeneration, cancer). With consideration of all organ systems (e.g., brain, heart, lung, gut, skeletal and smooth muscle, liver, pancreas, kidney, eye) and the interactions thereof, this Physiology Series will address the goals of resolving (1) Aging physiology and chronic disease progression (2) Examination of key cellular pathways as they relate to calcium, oxidative stress, and electrical signaling, and (3) how changes in plasma membrane produced by lipid peroxidation products can affect aging physiology, covering new research in the area of cell, human, plant and animal physiology.

Meet the Series Editor

Prof. Dr. Thomas Brzozowski works as a professor of Human Physiology and is currently a Chairman at the Department of Physiology and is V-Dean of the Medical Faculty at Jagiellonian University Medical College, Cracow, Poland. His primary area of interest is physiology and pathophysiology of the gastrointestinal (GI) tract, with a major focus on the mechanism of GI mucosal defense, protection, and ulcer healing. He was a postdoctoral NIH fellow at the University of California and the Gastroenterology VA Medical Center, Irvine, Long Beach, CA, USA, and at the Gastroenterology Clinics Erlangen-Nuremberg and Munster in Germany. He has published 290 original articles in some of the most prestigious scientific journals and seven book chapters on the pathophysiology of the GI tract, gastroprotection, ulcer healing, drug therapy of peptic ulcers, hormonal regulation of the gut, and inflammatory bowel disease.

Meet the Volume Editor

Prof. Dr. Hasan Basri İla received his Ph.D. from the Biology Department, Çukurova University, Institute of Sciences and Letters, Turkey. During his doctoral study, Dr. İla investigated the effects of a commonly used antibiotic on chromosome aberration and micronucleus formation by in vivo tests. He actively took responsibility for twenty-six national projects as a project leader and/or researcher and has given numerous poster and oral presentations at several international scientific conferences. He also lectures on biology, cytology, genetics, evolution, organelle genetics, and cancer genetics. Dr. İla has published several papers in internationally indexed journals and has written two book chapters and edited three books. He has one patent on natural pigment to his credit.

Contents

Preface

The force of natural selection tightly checks the adaptability of any living population to environmental conditions. On the other hand, the evolutionary success of a species in the phylogenetic path is measured by its flexibility against changing conditions. Like other intracellular components, peroxisomes also play an important role in the adaptation of the organism to its surrounding conditions. In eukaryotic organisms, the peroxisome, enclosed by a single membrane, is indispensable for lipid metabolism and free radical detoxification. The importance of peroxisome can be better grasped when the absolute harmful effects of free radicals as well as their newly discovered roles in intracellular signaling are evaluated holistically.

Despite extensive and rigorous scientific research, satisfactory information about the evolutionary origin of peroxisomes has yet to be elucidated. According to a theory-based argument, the ancestor of the peroxisome, which has an endosymbiotic origin, is a hydrogen-producing anaerobic prokaryotic organism that lost its DNA long ago or migrated to the host cell and dedicated itself to the endosymbiotic partnership with the eukaryotic cell earlier than the mitochondria and chloroplast. As evidence of the endosymbiotic origin of the peroxisome, the phylum Apicomplexa can be cited as a good example because this parasitic unicellular Sporozoa group contains mitochondria but not peroxisomes. However, the molecular composition and structural architecture of the peroxisomal membrane show homology to eukaryotic membranes. In addition, the molecular mobilization processes observed in the peroxisomal matrix protein import system and endoplasmic reticulum-associated protein degradation are interestingly similar.

Many diseases are related to peroxisome-induced free radical damage and intracellular signaling defect. As such, this book discusses prevention and sustainable treatment approaches for these diseases. Some of the highlights of the book are summarized below.

Disturbances in the reactive oxygen species (ROS) removal cascade of peroxisomes have important consequences for cell activities. The environment in which the cell interacts can either increase or decrease ROS damage. Today, the direct and indirect effects of natural and herbal supplements on the health of the organism are examined in a wide range. Communication disorders between the peroxisome and mitochondria can cause the oxidative balance to deteriorate in favor of reactive species. Peroxisomal dysfunction severely affects the morphological and molecular stability of the cell, resulting in apoptosis. Increased apoptosis is associated with many metabolic diseases that are often very difficult or impossible to treat. Molecules called peroxisome proliferator-activated receptors (PPARs) are nuclear receptor proteins that regulate the expression of genes that have important roles in many cellular processes, including tumorigenesis, as transcription factors. There are three subtypes of PPARs (PPAR-alpha, PPAR-beta/delta, and PPAR-gamma) found to be effective in the control of cancer metabolism steps. Disproportionate expression of any of these PPARs in the organism may lead to cell proliferation and abnormal survival (immortality) in various types of cancer. These molecules thus provide a potential alternative therapeutic possibility as both an agonist and antagonist target for cancer therapy. In this context, some natural supplements such as multifunctional PPAR modulators are likely to provide important instruments in the therapy of various types of cancer. Several recent studies have demonstrated that various phytochemicals, including phenolics, are associated with PPARs-mediated anticarcinogenicity. Considering the proposed mechanisms, it is envisaged that plant phenolics have both palliative and therapeutic opportunities in the fight against cancer. The ghrelin molecule is a hunger-stimulating peptide hormone produced by the stomach and pancreatic cells, and its levels rise before meals and then fall. This hormone has been shown to reduce food intake and body weight in rodents. Preproghrelin, the immature (unprocessed) form of ghrelin, is likely cleaved by a protease to synthesize a regulatory peptide, obestatin. One of the N-terminal fragments (Nt8U) of three different obestatins synthesized in the laboratory was found to be as active as obestatin in the experiments performed on Swiss albino mice. Obestatin and its fragment analog Nt8U are reported to upregulate glycerolipid metabolism and PPAR-gamma signaling and reduce fat accumulation in Swiss albino mice. Soon, computer-aided computational methods and advances in data processing analysis will be more widely used to answer the remaining questions about peroxisome regulation, mechanism, function, and its biogenesis in the context of healthy and pathological phenotypes. It is thought that advanced computational methods for the analysis of imaging data, protein structural modeling, proteomics, and genomics will come forward thanks to artificial intelligence (AI) such as machine learning and deep learning.

This book highlights the role of peroxisomes in the antioxidant mechanism as well as in the synthesis, maturation, and homeostatic regulation of some vital molecules. It also discusses diseases in humans caused by disruptions in peroxisome function and the modulatory effects of natural exogenous supplements on peroxisome activity.

Dr. Hasan Basri İla
Professor,
Faculty of Science and Letter,
Department of Biology,
Cukurova University,
Adana, Turkey

Section 1

Introduction

Chapter 1

Introductory Chapter: Peroxisome-Virus Interaction from SARS-CoV-2 Perspective

Hasan Basri İla

1. Introduction

Shannon Butalla and Melissa Gamble, parents of children affected by the peroxisomal disorder, took concrete steps in 2010 to establish the "Global Peroxisomal Disorders Foundation (GFPD)." The foundation has become an effective and complementary center providing research and family support for patients and their relatives struggling with rare peroxisomal diseases in the community. Peroxisomes, which are active in almost all our cells under normal conditions, are organelles that take an active role in lipid metabolism and render specific biochemical toxins harmless. Sabotaging the peroxisomes, which have supercritical initiatives for cell homeostasis, has irreparable consequences. Peroxisomal disorders include single-enzyme defects that affect a specific biochemical pathway and/or biogenesis defects that affect the entire peroxisome and therefore the organism. Many patients with peroxisome biogenesis defect (PBD) simultaneously lack the multiple functions of the peroxisome. These patients show a clinical gradient termed peroxisome biogenesis disorders-Zellweger Spectrum Disorders (PBD-ZSD). PBD-ZSD symptoms manifest in a wide range of patients. The disease symptoms can vary from mild as progressive hearing and vision problems after childhood to moderate-/high-level diseases including feeding problems, low muscle tone, and brain and liver disorders since infancy. Many patients with severely affected PBD-ZSD present with multiple medical problems, persistent seizures, developmental abnormalities, and even death in the neonatal period [1].

In patients with PBD-ZSD, autosomal recessive mutations have been found in a group of peroxisomal genes called peroxin (PEX), which plays a role in peroxisome biogenesis. In a typical PBD-ZSD family, each of the parents may be a carrier of a mutation in a gene such as PEX1. For example, the mutant PEX1 gene in both carrier parents does not affect the health of the carrier and does not produce severe disease for them. This is because there are two copies (diploidy) of each gene in each individual's genome, including PEX1. A healthy second copy can protect a carrier from the disease, but two carrier parents for the same gene have a 25% chance of transmitting the mutation to their children in each pregnancy. Because of their importance for peroxisome, the functions of PEX genes and their products, of which 37 have been discovered so far, are being studied intensively [2].

IntechOpen

2. Peroxisome virus interaction

2.1 Pandemic declaration and SARS-CoV-2

On March 11, 2020, it was recognized by the World Health Organization (WHO) that the epidemic caused by severe acute respiratory syndrome coronavirus 2 (SARS-CoV-2) (later named COVID-19) has become a pandemic. As of July 10, 2022, around 553 million confirmed cases and more than 6.3 million deaths have been reported worldwide due to COVID-19 [3, 4].

SARS-CoV-2 is a new strain of the coronavirus family that triggers and deepens health, social, and economic crises worldwide. Considering the effects of this virus on the organism, its interaction with the peroxisome is critical. SARS-CoV-2 infection in some human cell lines (Huh-7 and SK-N-SH) causes striking changes in peroxisome morphology. Since the peroxisomal membrane integrity is affected by the virus, peroxisomal matrix proteins leak into the cytosol and a marked reduction in the number of mature peroxisomes in the infected cells. Some proteins encoded by ORF14 of SARS-CoV-2 interact physically with human PEX14, which act as a membrane protein during matrix protein import and peroxisome biogenesis. Given the role of peroxisomes in innate immunity, the argument is getting stronger that SARS-CoV-2 can either directly target peroxisomes or is dangerous because it targets them. Therefore, a viral attack can cause the loss of function in the structural integrity, matrix protein content, and antiviral signaling of the peroxisome [5].

2.2 Interactions of some viruses with peroxisomes

A newly discovered mechanism in human immunodeficiency virus (HIV) patients (AIDS) has offered hope as a different approach for treating COVID-19. More recently, the steps of HIV attack on peroxisomes, which regulate many cellular mechanisms, including the immune system, have been demonstrated. It has been demonstrated that some viruses, including West Nile virus (WNV) and ZIKA virus (ZIKV), have different instruments to inhibit the host-cell interferon response, such as reducing peroxisomes. In fact, all viruses have ways of blocking the host-interferon response. This situation is consistent with the concept that peroxisomes are an important target for viral infections. At this point, researchers evaluated that SARS-CoV-2 could also attack peroxisomes and started testing drugs that increase peroxisome activity against the virus in cell cultures. Recently, a group of scientists reported the mechanistic basis for why many HIV patients suffer from premature aging, lipodystrophy (change in the body fat metabolism cascade), and several diseases. The team identified four microRNAs with increased expression in the brains of HIV patients with dementia. All four microRNAs increased in the brains of these patients appeared to target the peroxisome biogenesis pathway. These increased expression molecules downregulate protein expression required to form peroxisomes. On the other hand, it has been shown that increased peroxisomes by manipulating appropriate genes can inhibit ZIKV replication. Tests of drugs that increase peroxisomes are ongoing to evaluate the success against SARS-CoV-2 and HIV. Many of the agents that researchers have tried have been approved as cancer drugs. During these investigations, it was discovered that a pathway that inhibits peroxisome formation was targeted. Despite possible indications for drug use, it is noted that some peroxisome boosters can be taken orally and have low side effects [6].

3. The plasmalogen importance for viral defense

The cellular functions of plasmalogens, which are abundant in many mammalian tissues, have not yet been fully elucidated. However, it is evaluated that they can protect cells against the harmful effects of endogenous reactive oxygen species. It is also thought to act as signaling molecules and modulators in the membrane [7–9]. Plasmalogen biosynthesis requires functional peroxisomes that are oxidative organelles in the cell. Peroxisomes are important host organelles where certain viral replication can take place [10]. Plasmalogen deficiency observed in cardiometabolic and multiple neurodegenerative diseases may predispose humans to SARS-CoV-2 and other similar viral infections. On the other hand, increased plasmalogen levels were detected in humans infected with some viruses (ZIKV, HBV, and HIV) and in the virion lipidome of cytomegalovirus [11–14]. Increased plasmalogen levels revealed a strong association between the ZIKV lifecycle and host peroxisomes. This finding scientifically explains the upregulation of resident plasmalogens and peroxisome-mediated lipidome changes in the serum of patients infected with ZIKV [14, 15]. Meanwhile, it has been reported that the plasmalogen phosphatidylcholine molecule plays an important role in influenza virus infection [16].

The infection response of the host cell can be used as the basic determinant in the pathogenesis of any infectious disease, including the COVID-19 pandemic. The lipid composition of the host cell plasma membrane has a decisive role in the life cycle of a virus, because the first step of the infection process is the entry of the virus into the cell by crossing the membrane. In viral infections in a cell, peroxisomes act as vital immune signaling foci and assist the host by regulating antiviral signaling [10]. Peroxisomes are critical organelles that act as a double-edged sword in the viral infection process. Since this cellular organelle, which both harbors and kills the pathogen, has a primary role in host antiviral defense, it can transform into a useful apparatus that serves viral replication through critical manipulations [17]. This interaction in favor of the virus brings to mind the famous aphorism of nineteenth-century German philosopher Friedrich Nietzsche "What doesn't kill me, makes me stronger."

With the discovery of mitochondrial antiviral signaling (MAVS), an innate host immune response, mitochondria were recognized as an important subcellular signaling center. Initially, MAVS, which produced a rapid antiviral reaction, was thought to be specific to mitochondria only. However, considering the known roles of peroxisomes in detoxification, the identification of peroxisomal MAVS (or PAVS) has revealed the importance of peroxisome in host defense as an antiviral signaling organelle. This finding was supported by the discovery of increased host peroxisome biogenesis during human cytomegalovirus (HCMV) and herpes simplex type 1 (HSV-1) infections [18]. The synthesis of some vital molecules, which are essential for viral penetration and the success of the virus life cycle, only in the peroxisome is the most important link in the chain. These molecules are plasmalogens and some cellular lipids such as docosahexaenoic acid, a very long chain omega-3 fatty acid. The units that synthesize plasmalogen and some lipids are peroxisomes, which are indispensable for the construction of viral envelopes, modulation of host cholesterol homeostasis, and maintenance of virus-host balance during infection. It is therefore not surprising that peroxisomes are an attractive candidate for cellular remodeling during some viral infections. Consistent with this proposition, it has been suggested that increased plasmalogen in HCMV virions [12] and alterations in peroxisomal lipid metabolism may be a general characteristic of enveloped virus infections. Plasmalogen and some lipids are essential components of many enveloped viruses, including HCMV and influenza. Therefore, it is considered that this argument may also be valid for SARS-CoV-2. However, further lipidomic analysis is required in COVID-19 samples for a more detailed projection [19].

Macrophages, professional phagocytes of the host immune system, can detect and clear invading pathogens such as viruses and damaged cells. Plasmalogen deficiency in macrophages is associated with a reduced ability to phagocytosis. This situation is significantly reversed when cells are exposed to lysophosphatidylethanolamine plasmalogen [20]. Similarly, restoration of plasmalogen levels leads to increases in the number and size of lipid microdomains in the membranes of macrophages. Therefore, exogenous plasmalogen administration is likely to be adopted as an innovative strategy for optimizing macrophage function [19]. According to the results of a comprehensive bioinformatics study on macrophage differentiation, the plasmalogen phosphatidylethanolamine (PE) molecule is a biomarker of immune system activation. As an interesting finding, a significant decrease in plasmalogen levels was observed in obese subjects [21]. In light of these findings, the potential link between host plasmalogen dysregulation and the high morbidity and mortality levels observed in COVID-19 patients is considered significant. Again, a strong correlation is observed between decreased plasmalogen levels and a number of pathological conditions, including neurodegenerative and cardiometabolic disorders, as well as severe COVID due to coronavirus infection. In coronavirus-induced lipidome patterns, irregular plasmalogen levels in the infected patients are of interest. This finding indicates that the plasmalogen molecule is among the key lipids in potentially modulating viral infection [19].

4. The host cell reprogramming by SARS-CoV-2

SARS-CoV-2 replication alters the morphology, number, and function of many cellular structures. Peroxisome accumulation was observed in regions containing double-membrane vesicle (DMV) clusters formed by the endoplasmic reticulum (ER). This finding was confirmed by detecting a marked increase in peroxisome-associated protein, PMP70, in infected cells by confocal microscopy and Western blot analysis. The convergence between the peroxisome and the viral RNA replication site serves to protect viral RNA from oxidative damage [10] or to establish a signaling platform that generates a cytokine response [22].

It has been discovered that many cellular components other than the ER and peroxisomes are also remodeled in cells infected with SARS-CoV-2. It is noteworthy that this transformation is accompanied by decreased mitochondrial ATP synthase, swollen cristae, and matrix condensation. The related observation is consistent with reduced oxidative phosphorylation in cells infected with SARS-CoV-2 and transcriptional changes pointing to virus-induced metabolic reprogramming [23]. The virus in the infected cell manages dynamic changes in the biosynthesis and utilization of macromolecules such as glucose, nucleotides, fatty acids, and amino acids to expand the maneuvering space. These changes are directly related to the success of viral infection [24].

5. Inference

As can be seen from the results of the studies discussed above, there is a multilayered, dynamic, and complex interaction between peroxisome health and the virus life cycle. The coincidental and rapid evolution of viruses in the direction of manipulating the peroxisomal pathway at any stage during the infection process may lead to dead ends or irreparable damage to the host cell defense system. In general, the possibility that extraordinary viruses such as SARS-CoV-2 may favor people with fragile peroxisomes for genetic or environmental reasons should be seriously considered. In this context, handling new insights such as peroxisome health in the control of viral pandemics will provide an important achievement in the fight against the disease.

Author details

Hasan Basri İla
Faculty of Science and Letter, Department of Biology, Çukurova University, Adana, Turkey

*Address all correspondence to: ila@cu.edu.tr; hasanbasriila@gmail.com

References

[1] GFPD. The Global Foundation for Peroxisomal Disorders [online]. 2022. Available from: https://thegfpd.org/. [Accessed: July 21, 2022]

[2] Jansen RLM, Santana-Molina C, van den Noort M, Devos DP, van der Klei IJ. Comparative genomics of peroxisome biogenesis proteins: Making sense of the PEX proteins. Frontiers in Cell and Developmental Biology. 2021;**9**:1229

[3] WHO. WHO Director-General's opening remarks at the media briefing on COVID-19-11 March 2020 [online]. 2020. Available from: https://www.who.int/director-general/speeches/detail/who-director-general-s-opening-remarks-at-the-media-briefing-on-covid-19---11-march-2020. [Accessed: July 22, 2022]

[4] WHO. Weekly Epidemiological Update on COVID-19-13 July 2022 [online]. 2022. Available from: https://www.who.int/publications/m/item/weekly-epidemiological-update-on-covid-19---13-july-2022. [Accessed: July 20, 2022]

[5] Knoblach B, Ishida R, Hobman TC, Rachubinski RA. Peroxisomes exhibit compromised structure and matrix protein content in SARS-CoV-2-infected cells. Molecular Biology of the Cell. 2021;**32**(14):1273-1282

[6] Rutherford G. Discovery of Mechanism behind HIV Infections Could Also Hold Key to COVID-19: U of A Researchers | Folio [Online]. Canada: University of Alberta; Available from: https://www.ualberta.ca/folio/2020/07/discovery-of-mechanism-behind-hiv-infections-could-also-hold-key-to-covid-19-u-of-a-researchers.html; 2020 [Accessed: July 20, 2022]

[7] Nagan N, Zoeller RA. Plasmalogens: Biosynthesis and functions. Progress in Lipid Research. 2001;**40**(3):199-229

[8] Gorgas K, Teigler A, Komljenovic D, Just WW. The ether lipid-deficient mouse: Tracking down plasmalogen functions. Biochimica et Biophysica Acta (BBA) - Molecular Cell Research. 2006;**1763**(12):1511-1526

[9] Moser AB, Steinberg SJ, Watkins PA, Moser HW, Ramaswamy K, Siegmund KD, et al. Human and great ape red blood cells differ in plasmalogen levels and composition. Lipids in Health and Disease. 2011;**10**:101

[10] Cook KC, Moreno JA, Jean Beltran PM, Cristea IM. Peroxisome plasticity at the virus-host Interface. Trends in Microbiology. 2019;**27**(11):906-914

[11] Brügger B, Glass B, Haberkant P, Leibrecht I, Wieland FT, Kräusslich HG. The HIV lipidome: A raft with an unusual composition. Proceedings of the National Academy of Sciences of the United States of America. 2006;**103**(8):2641-2646

[12] Liu STH, Sharon-Friling R, Ivanova P, Milne SB, Myers DS, Rabinowitz JD, et al. Synaptic vesicle-like lipidome of human cytomegalovirus virions reveals a role for SNARE machinery in virion egress. Proceedings of the National Academy of Sciences of the United States of America. 2011;**108**(31):12869-12874

[13] Schoeman JC, Hou J, Harms AC, Vreeken RJ, Berger R, Hankemeier T, et al. Metabolic characterization of the natural progression of chronic hepatitis B. Genome Medicine. 2016;**8**(1):1-13

[14] Queiroz A, Pinto IFD, Lima M, Giovanetti M, de Jesus J.G, Xavier J, et al. Lipidomic analysis reveals serum alteration of plasmalogens in patients infected with ZIKA virus. Frontiers in Microbiology. 2019;**10**(APR):753

[15] Martín-Acebes MA, Oya NJ de and Saiz JC. Lipid metabolism as a source of druggable targets for antiviral discovery against Zika and other Flaviviruses. Pharmaceuticals (Basel, Switzerland). 2019;**12**(2):97

[16] Tanner LB, Chng C, Guan XL, Lei Z, Rozen SG, Wenk MR. Lipidomics identifies a requirement for peroxisomal function during influenza virus replication. Journal of Lipid Research. 2014;**55**(7):1357-1365

[17] Karnati S, Baumgart-Vogt E. Peroxisomes in mouse and human lung: Their involvement in pulmonary lipid metabolism. Histochemistry and Cell Biology. 2008;**130**(4):719-740

[18] Jean Beltran PM, Cook KC, Hashimoto Y, Galitzine C, Murray LA, Vitek O, et al. Infection-induced peroxisome biogenesis is a metabolic strategy for herpesvirus replication. Cell Host & Microbe. 2018;**24**(4):526-541.e7

[19] Deng Y, Angelova A. Coronavirus-induced host cubic membranes and lipid-related antiviral therapies: A focus on bioactive plasmalogens. Frontiers in Cell and Developmental Biology. 2021;**9**:630242

[20] Rubio JM, Astudillo AM, Casas J, Balboa MA, Balsinde J. Regulation of phagocytosis in macrophages by membrane ethanolamine Plasmalogens. Frontiers in Immunology. 2018;**9**(JUL):1723

[21] Wallner S, Grandl M, Konovalova T, Sigrüner A, Kopf T, Peer M, et al. Monocyte to macrophage differentiation goes along with modulation of the plasmalogen pattern through transcriptional regulation. PLoS One. 2014;**9**(4):e94102

[22] Dixit E, Boulant S, Zhang Y, Lee ASY, Odendall C, Shum B, et al. Peroxisomes are signaling platforms for antiviral innate immunity. Cell. 2010;**141**(4):668-681

[23] Neufeldt CJ, Cerikan B, Cortese M, Frankish J, Lee J-Y, Plociennikowska A, et al. SARS-CoV-2 infection induces a pro-inflammatory cytokine response through cGAS-STING and NF-κB. bioRxiv. 2020. 2020.07.21.212639

[24] Fantacuzzi M, Amoroso R, Ammazzalorso A. PPAR ligands induce antiviral effects targeting perturbed lipid metabolism during SARS-CoV-2, HCV, and HCMV infection. Biology. 2022;**11**(1):114

Chapter 2

The Mystery of Peroxisomes

Hasan Basri İla

Abstract

According to the evolutionary perspective, an organism must manage and optimize organized complexity effectively to achieve a strong adaptation. Within the scope of sustainable homeostasis, the subcellular components of the organism must strictly comply with the principle of minimum error and maximum efficiency in coordination. Advanced defense systems are evolution's greatest gift to the cell. One of the most important components of cellular defense systems is the antioxidant defense. When it comes to antioxidant defense, the first thing that comes to mind is the peroxisome organelle, because the peroxisome is a cytoplasmic organelle surrounded by a single membrane in which the very important enzyme, catalase, is localized. Furthermore, the role of this organelle in vital processes, such as lipid metabolism, antimicrobial defense, and intracellular signaling, is undeniable. In this chapter, attention has been tried on the mysteries related to peroxisome by performing a wide literature review. The chapter covers topics such as peroxisome production, targeted protein transport, roles in the oxidative mechanism, relationship with diseases, and mitochondria interaction. This chapter, which highlights the polygenic formation and pleiotropic features of peroxisome, will provide an important future projection for curious researchers and medical doctors seeking innovative treatment strategies.

Keywords: peroxisome, oxidative stress, mitochondria, protein import, cell differentiation, cancer

1. Introduction

The major oxidation event worldwide occurred about 2.45 billion (B) years ago. However, there is heated debate that an earlier and failed oxygenation event may have occurred about 3.2 B years ago. However, the absence of sedimentary rock deposits stained red with iron oxide in stratigraphic units older than 2.5–2.0 B years is considered to be the most convincing evidence of non-oxygenation [1]. The study of sulfur assets on earth shows that between 2.45 and 2.09 B years ago a change occurred in the sulfur cycle. Atmospheric reactions at that time and the partial pressure of atmospheric oxygen (O_2) also played a role in determining the oxidation state of sulfur in the earth's crust. The findings indicate that the atmospheric oxygen partial pressures were low at that time and their roles in oxidative decomposition, microbial oxidation, and sulfur reduction were minimal [2].

According to the consensus accepted by the majority of scientists, the axis of the evolution of living things changed markedly as atmospheric oxygen started to increase from 2.5 to 2 B years ago. Since the first atmospheric oxygen appeared, there have been drastic fluctuations in oxygen concentration. After a significant increase 375–275 million years ago, it decreased slightly and reached today's level (21%) [3]. Mitochondria allowed the development of eukaryotes, thanks to the exploitation of this dangerous and highly reactive element, O_2 (oxygen respiration). Reactive oxygen species (ROS), such as superoxide, hydrogen peroxide, and hydroxyl ($O_2\cdot-$, H_2O_2, and •OH), some of which are quite dangerous, have widely entered cellular life due to the oxidation and reduction (redox) reactions that take place in aerobic respiration. In addition to being a reactive species, H_2O_2, one of these ROS, also functions as a signal molecule in the intracellular metabolic pathways [4].

The primary source of H_2O_2 production in the cell is mitochondria [5]. Complex I, one of the components of the electron transport chain in mitochondria, firstly causes the production of superoxide ($O_2\cdot-$), which is then converted to H_2O_2 by being dismutated by manganese superoxide dismutase (Mn-SOD) induced by NF-κB activation [4]. Peroxisomes, which play a central role in lipid metabolism, owe their name to activities that produce and scavenge hydrogen peroxide and are eukaryotic organs that also play key roles in the conversion of reactive oxygen species. Peroxisomes are responsible for the catabolism of very-long-chain fatty acids, branched-chain fatty acids, bile acid intermediates, D-amino acids, and polyamines, and the reduction of reactive oxygen species, especially hydrogen peroxide [6]. In addition, peroxisomes play a role in the biosynthesis of ether phospholipid plasmalogens, which are critical for the normal function of organs such as the mammalian brain, heart, and lungs. In humans, plasmalogens constitute approximately 18% of the total phospholipid mass and show a cell- and tissue-specific distribution [7]. Plasmalogens (1–0-alk-1′-enyl-2-acyl glycerophospholipids) constitute a special class of phospholipids characterized by the presence of a vinyl-ether bond at the sn-1 position [8].

In addition to the presence in the cytosol of two enzymes in the pentose phosphate pathway (glucose-6-phosphate dehydrogenase and 6-Phosphogluconate dehydrogenase), which are important for energy metabolism (NADPH), the same enzymes are also found in intact peroxisomes secretly and a part of the total activity is associated with these enzymes located peroxisomal [9]. This finding is an indication of the tight relationship and strong interaction between peroxisome and other subcellular compartments.

2. Peroxisome

Peroxisomes were identified in 1960 as part of the pioneering work of Christian René de Duve, who developed cell lysis techniques. De Duve's method separates organelles according to their precipitation and density characteristics, and it has been determined that peroxisomes are more intensive in the cell than other organelles [10]. Peroxisomes, which are versatile (pleiotropic) organelles, constantly adapt (plasticity) to current environmental conditions. The number of this organelle is rapidly adjusted in response to cyclical changes or various stimuli to maintain a steady state in the cell. Although the mystery of peroxisomes is still not fully resolved, it can be generalized that it is an organelle involved in (anti)oxidation and some synthesis pathways. Although morphologically similar to lysosomes, the structure is matured by the transport of proteins synthesized in free ribosomes and then functionalized to peroxisomes. At this point, the debate over whether peroxisomes proliferate by division or by *de novo* synthesis draws attention. Peroxisomes do not have their genome but replicate similar to mitochondria and chloroplasts. Peroxisomes contain at least 50 different enzymes involved in various biochemical pathways in different cell types. Initially, they were described as organelles that carry out oxidation reactions that lead to the production of hydrogen peroxide. On the other hand, since hydrogen peroxide is harmful to the cell, it is vital to render it harmless. For this purpose, the peroxisome either breaks down hydrogen peroxide directly into water and oxygen via its catalase enzyme.

$$2H_2O_2 \Rightarrow 2H_2O + O_2 \tag{1}$$

Various substrates, including uric acid, amino acids, and fatty acids, are broken down by oxidative reactions that take place in peroxisomes. The oxidation of fatty acids is an important example as it provides a great source of metabolic energy. Fatty acid oxidation in peroxisomes is accompanied by the production of hydrogen peroxide (H_2O_2). Although fatty acids are oxidized in both mitochondria and peroxisomes in animal cells, fatty acid oxidation in yeast and plants is limited to peroxisomes. In animal cells, cholesterol and dolichol are synthesized in peroxisomes as well as in the ER. Peroxisomes in the liver are also involved in the synthesis of bile acids derived from cholesterol. In addition, peroxisomes regulate sterol regulatory element-binding proteins (SREBPs) and SREBP cleavage-activating protein (SCAP) traffic that activates this protein. This regulation is of functional importance for maintaining cholesterol homeostasis and efficient cholesterol synthesis. In addition, peroxisomes contain enzymes necessary for the synthesis of plasmalogens, a member of the glycerophospholipid family. In the chemical structure of plasmalogen, one of the hydrocarbon chains is attached to glycerol with an ether bond instead of an ester bond. Although not widely observed in other tissues, plasmalogens are important membrane components in some tissue cells, especially in the heart and brain [11, 12].

2.1 Glyoxysome

Special peroxisomes found especially in seed plants as well as filamentous fungi are called glyoxysomes. Peroxisomes play two particularly important roles in plants. First, peroxisomes are responsible for metabolizing fatty acids stored in seeds into critical carbohydrates to provide growth energy and raw materials to the germinating plant. This process, called the glyoxylate cycle, is a variation of the citric acid cycle that occurs through a series of reactions. Second, peroxisomes in plant leaves are involved in the metabolism of the by-product formed in the photorespiration process [12].

2.2 Reproduction and growth of peroxisomes

As noted earlier, the peroxisome is similar in form to mitochondria and chloroplasts in terms of some fundamental metabolism. Structural and enzymatic proteins of peroxisomes are synthesized by cytosolic free ribosomes (sometimes ER polysome) and shipped specifically to the peroxisome. Similarly, phospholipids are transferred from major synthesis sites in the ER to peroxisomes via phospholipid transfer proteins (PLTP). With the import of protein and phospholipids, peroxisomes grow and then divide to form new peroxisomes.

Thirty-seven peroxisome biogenesis factor proteins (known as Peroxin or PEX) that are involved in peroxisome biogenesis and proliferation identified so far are encoded by PEX genes. However, not all eukaryotes contain peroxisomes, so there are several PEX proteins involved in possible alternative processes in organisms lacking peroxisomes. Initial data revealed that several protist species (*Cryptosporidium parvum, Theileria annulata, Babesia bovis, Monosiga brevicollis, Plasmodium falciparum, Blastocystis hominis,* and *Entamoeba histolytica*) lack most PEX proteins [13]. Some eukaryotes, such as anaerobic protists, plasmodium, and parasitic platyhelminths, lost their peroxisomes in the evolutionary process. Surprisingly, *Oikopleura dioica*, a free-living pelagic tunic in the oxygen-containing niches of marine waters, contains no peroxisomal gene in its genome. It was also found that the putative peroxisomal enzyme set was considerably reduced and none contained a predicted peroxisomal targeting signal (PTS). It has been shown that several metazoan lineages, such as *O. dioica*, independently lost peroxisomes showing that peroxisome loss is not only related to adaptation to anaerobic habitats and/or parasitic lifestyle [14].

PEX genes are required for peroxisome growth and division. Overexpression of the PEX11 gene causes peroxisome proliferation, while its deletion leads to the enlargement and reduction of peroxisomes. While Pex11 causes membrane tubulation *in vitro*, the cytoskeleton also contributes to this situation likely [15].

2.3 Protein transport into the peroxisome

According to the peroxisome database, a complete peroxisomal proteome is encoded by 61 genes in *Saccharomyces cerevisiae* and 85 genes in *Homo sapiens* [16]. Targeting of proteins into the peroxisome is achieved by at least two conserved pathways from yeasts to humans. Most proteins are targeted to peroxisomes by the simple amino acid sequence (Ser-Lys-Leu tripeptide) at their carboxy terminus (peroxisome targeting signal 1 or PTS1) [12]. However, there is a different source suggesting that amino acids in this signal (PTS1) show polymorphism as (Cys/Ala/Ser)-(Lys/Arg/His)-(Leu/Ile) [17]. Other peroxisomal proteins are targeted to the organelle by another signal peptide (PTS2) located at the amino terminus. Although there are multiple variants of PTS2, it consists of the nine amino acid consensus nonapeptide [(-Arg/Lys) (Leu/Val/Ile) XXXXX (His/Gln) (Leu/Ala)] sequence (X in this sequence represents any amino acid) showing amino acid polymorphism. However, some proteins may be targeted by alternative signals that are not yet well-defined [18]. PTS1 and PTS2 are recognized by different receptors and then transferred to a translocation complex in the peroxisome membrane. Cytosolic heat shock protein 70 (Hsp70) has been associated with protein import into peroxisomes, but the possible role of molecular chaperones in peroxisomes is unclear [12]. According to new data, Hsp70s mediates the delivery of components such as the ubiquitin E3 ligase Stub1 to the damaged peroxisome. In this way, peroxisome may promote autophagic degradation (pexophagy) of itself due to oxidative stress. In summary, Hsp70s are involved in the negative regulation of peroxisomes [19].

Three receptors, namely Pex5p, Pex7p, and Pex19p have been identified so far that recognize various types of PTS ligands. Pex5p functions as a cyclic transport receptor for proteins carrying newly synthesized PTS1 or not carrying any targeting signals. Pex5p recognizes and binds proteins to be transported in the cytosol and transports them to the peroxisomal matrix. This process is based on complex and transient protein-protein interactions involving cargo recognition, cargo-loaded receptor docking to the peroxisomal membrane, cargo translocation, and release. To fulfill these functions, all Pex5p orthologs contain multiple flexible segments. It contains two clusters of three tetratricopeptide repeat (TPR) motifs in the C-terminal part of the molecule, connected by a flexible joint region. This construct creates a single binding site for PTS1. The N-terminal part of the molecule has multiple diaromatic pentapeptide motifs (usually expressed as TrpXXXPhe/Tyr). The number of these peptide motifs varies between species, with two in *S. cerevisiae* and nine in *Arabidopsis thaliana*. Duly, both the C- and N-terminal domains of Pex5p undergo significant conformational transitions after binding to their ligands.

The Pex7p molecule is a soluble protein that serves as a targeting signal recognition factor for newly synthesized PTS2 proteins. It exhibits a cytosolic and intraperoxisomal distribution pattern and can be repeatedly translocated inside and outside of the organelle. All Pex7p orthologs are characterized by the presence of six WD40[1] repeat motifs, which together with a distinct N-terminal region are predicted to form a seven-bladed β-propeller-like structure. Mutations that affect the conformation of this structure almost always abolish activity. PTS2's receptor, Pex7p, recruits cofactors for its function in peroxisomal protein uptake. These co-factors termed "PTS2 coreceptors," are species-specific and include an insertion form similar to Pex5pL, the long isoform of Pex5p in mammals. All these proteins are largely cytosolic and show a low overall similarity between their primer sequences.

The Pex19p molecule is a predominantly cytosolic, partially peroxisome localized multifunctional protein and plays a central role in the early steps of peroxisomal membrane synthesis. It has been determined that this Peroxin can (i) bind newly synthesized distinct peroxisome membrane proteins (PMP) in the cytosol, (ii) keep these PMPs in a competent conformation for attachment to the membrane, (iii) transport them to the peroxisomal membrane, and (iv) return to the cytosol with a shuttle-like movement. Because of this feature, the suggestion that Pex19p functions as a chaperone and soluble transport receptor for "class I" PMPs is strengthened. Members of this PMP class contain common Pex19p binding motifs that are an integral part of targeting signals (PTSs). Interestingly, many PMPs contain multiple Pex19p binding motifs, but not all of these regions can directly bind to PTS [20].

Some peroxisome membrane proteins are similarly synthesized by cytosolic ribosomes and targeted to the peroxisomal membrane by different internal signals. However, the expression of some peroxisomal membrane proteins in membrane-bound polysomes of the endoplasmic reticulum and their transport to peroxisomes point to the important role of the endoplasmic reticulum for peroxisome integrity [12].

Specific protein complexes exist that act as docking sites on the peroxisomal membrane for cargo-loaded PTS receptors. Currently, two complexes have been identified that perform this task: one recognizes the incoming Pex5p- and Pex7p-cargo complexes, and the other recognizes the binding of the Pex19p-cargo complex. Here, it was determined that Pex5p Peroxin showed a higher affinity for Pex14p and the amount of peroxisome-associated Pex5p was proportional to the amount of Pex14p. In addition, it has long been known that in many species deficient in Pex3p, Pex16p, or Pex19p, the peroxisomal membrane structures of cells lack integrity. This observation led to the hypothesis that these Peroxins are essential for peroxisome membrane biogenesis. It has been shown that Pex8p from *Pichia pastoris*, a

---

[1] It is a short structural motif of about 40 amino acids, usually ending with a tryptophan-aspartic acid dipeptide. Successive copies of these repeats fold together to form a type of circular solenoid protein called the WD40 domain.

methylotrophic yeast species, requires only Pex5p and Pex14p for PTS1-dependent transfer to peroxisomes. It has been reported that the diameter of the ion-gated channel formed in studies with proteoliposomes may expand when it encounters cytosolic Pex5p-cargo complexes. The Pex5p/Pex14p proteins herein are regulatory elements of the matrix protein transition pore in the peroxisome membrane. In summary, there is currently substantial evidence that some PMPs migrate through the ER to peroxisomes, while others separate directly from the cytosol into these organelles. There is also no consensus on whether these processes require an energy source. While the interactions between PTS receptors Pex5p, Pex7p, and Pex19p and their cargo proteins are relatively well-characterized, little information is available about how the cargo, carrier, and receptor complexes are separated during the delivery of cargo to its destination. It was initially suggested that Pex8p, an intraperoxisomal protein containing both PTS1 and PTS2 signaling, may act as a PTS1 receptor-cargo release factor. However, since Pex8p is only found in fungi (such a mechanism is not active in higher eukaryotes), it has been suggested that dissociation of the Pex5p-PTS1 complex may be mediated by a change in pH. According to this hypothesis, Pex5p exists in different oligomeric conformations in the cell, and these structures change with pH. Pex5p is in the monomeric form at pH 6.0 and evolves into a tetrameric form at pH 7.2. While PTS1 peptides bind to tetrameric Pex5p predominantly under slightly alkaline conditions, due to the slight acidity of the peroxisomal matrix, the structure changes to monomeric conformation, resulting in the release of cargo. However, factors other than pH may also trigger the cargo release step at a higher level. The addition of single or multiple ubiquitin to conserved cysteine and lysine residues at the N-terminus of Pex5p, Pex18p, Pex20p, or Pex21p molecules can be expressed as a dissociation factor other than pH. Conjugation of ubiquitin-like molecules to a protein requires the concerted action of an ATP-requiring ubiquitin-activating enzyme (E1), a ubiquitin-conjugating enzyme (E2), and a ubiquitin-ligase (E3). It should be noted that mutations in any of these proteins will result in a defect in peroxisomal matrix protein import. For sustainable protein uptake (import) cycle, (ubiquitinated) PTS (co)receptors must be returned to the cytosol with (by) shuttle-like movement. Selective elimination of unnecessary and dysfunctional peroxisomes is required to regulate peroxisome function properly and limit damage during cellular aging. Biochemical and genetic studies in different organisms have shown that the degradation used to eliminate peroxisome can occur through at least three different mechanisms. These are macropexophagy, micropexophagy, and 15-lipoxygenase-mediated autolysis. More than 35 autophagy-related (ATG) genes have been identified to date. The proteins encoded by these genes, collectively called Atg protein genes, are essential for selective and nonselective autophagy pathways. Interestingly, all these pathways have been stated to require a core molecular machinery conserved from yeast to humans [20].

3. Effect on oxidative stress

Peroxisomes, which emerged as the product of the selective strategy of evolution seeking a way out, break down various substrates including amino acids, fatty acids, and uric acid in the cell, as is commonly known, through oxidative reactions, while reactive byproducts are also produced. Peroxisomes both render harmless the reactive products formed by redox reactions and are important actors in the biosynthesis of a phospholipid, dolichol, plasmalogen, cholesterol, and bile acid [12].

The radical scavenging function of peroxisome has been emphasized in the historical process from its discovery to the present. However, it produces radicals for various reasons in metabolic pathways, and then it renders them harmless in the cell. Reactive oxygen species (ROS) take an active role in the natural biosynthetic cascade, both as an intracellular signaling molecule and for the breaking or rebuilding of existing molecular bonds, due to their effect on electrical charge distribution [21, 22]. Interestingly, highly dangerous reactive molecules take an active mission in cell molecular architecture. On the other hand, the fact that peroxisomes contain antioxidant enzymes to eliminate the damage of radical species is related to the dual activity of the peroxisome. This reality brings to mind the struggle of Ahura Mazda - Ahriman in the Zoroastrian belief. Peroxisomes interact and cooperate closely with other cell organelles (mitochondria and chloroplasts) for the optimization of cellular homeostasis. In addition, due to the excellent plasticity of the peroxisomes, their number, shape, and protein contents are dynamically regulated in response to changing cellular and environmental conditions. Although the peroxisomal β-oxidation pathway is similar to the mitochondrial pathway, the related enzyme sets in the reaction are different. Unlike mitochondria, ATP is not released in the fatty acid breakdown in peroxisomes. The fatty acid is activated by the fatty acyl CoA synthetase in the first step in the degradation process. First, a double bond is added to the beta position of the fatty acyl-CoA ester. This step, catalyzed by FAD-containing acyl-CoA oxidase, generates hydrogen peroxide by transferring hydrogen atoms to molecular oxygen. Constitutive activation of peroxisome proliferator-activated receptor alpha (PPARα) occurs when fatty acids accumulate due to defective degradation. PPAR activation upregulates the levels of peroxisomal and mitochondrial β-oxidation enzymes that can lead to oxidative stress. In addition, complex forms of transition elements such as iron and copper are abundant in the peroxisome. Therefore, some toxic xenobiotics (pyrimidines, dialuric acid, divicine, isouramil, etc.) cause the iron release from iron-binding proteins such as ferritin. In the Fenton reaction, it has been shown that in the presence of ascorbate, histidine, or ADP, Fe^{2+} induces lipid peroxidation by breaking the H_2O_2 molecule into more dangerous hydroxyl (OH) radicals. On the other hand, peroxisome contains a number of antioxidant enzymes to prevent oxidative damage of hydrogen peroxide itself and free radicals derived from it. Hydrogen peroxide is cleared by catalase and glutathione peroxidase. Again, superoxide anions MnSOD and CuSOD are rendered harmless by converting them to hydrogen peroxide by Zn-SOD. Along with the last two enzymes, catalase and glutathione peroxidase are mainly found in mitochondria and cytosol, but these enzymes have also been reported to be present in peroxisomes. However, questions about how peroxisomal response signals are created have only recently begun to be answered. Located in the peroxisome membrane and a central component of the protein import process, Pex14 is first phosphorylated in mammalian cells in response to oxidative stresses such as H_2O_2. In this process, H_2O_2-induced phosphorylation of the Ser232 residue of Pex14 suppresses the import of catalase *in vivo*, and this phosphorylation selectively impairs the *in vitro* interaction of catalase with the Pex14-Pex5 complex. It has been found that this has an effect on cytosolic and peroxisomal catalase levels [23–25].

4. Peroxisome-related illness or genetic disorders

Enzymatic antioxidants in the peroxisome like catalase play important roles. Some *de facto* conditions in the enzymes and other auxiliary factors that take part in the conversion of superoxide to the reduction stages to more stable molecules and the breakdown of H_2O_2 into water and oxygen will cause the creation of the •OH radical. The •OH, the hazardous premature product of H_2O_2 is a free radical that carries a serious risk for all organic molecules. It attacks all forms of molecules, including nucleic acids (DNA and RNA), and changes their molecular conformation. H_2O_2, which is also a known DNA disrupter, has taken its place in test protocols as a known mutagen in some genetic tests (e.g., Comet assay). There is no doubt the costs of the weaknesses that may arise in antioxidant enzymes will be heavy for the cell. However, a serious paradox stands here. Accordingly, do increased genetic defects (mutations) lead to oxidation, or do increased oxidations lead to mutations? This situation is similar to a popular Ancient Greek paradox, "*Which came first, the chicken or the egg*?" The undeniable fact is that increased ROS causes mutations, while increased mutations produce more ROS. It is like the Ouroboros[2] phenomenon. Acatalasemia with inherited catalase activity deficiency was initially considered an asymptomatic disorder. Decreases in the extracellular hydrogen peroxide (H_2O_2) removal capacity of catalase-deficient tissues have been determined in animal models. In patients with catalase deficiency, H_2O_2 can cause methemoglobinemia[3]. The high (18.5%) prevalence of diabetes and the onset of the disease 10 years earlier in individuals with hereditary catalase deficiency may be attributed to oxidative damage to oxidant-sensitive, insulin-producing pancreatic beta cells. Oxidative stress and aging-related diseases were diagnosed in 97 of 114 acatalasemias. Oxidative stress due to catalase deficiency alone may contribute to the onset of diabetes and may also be one of the causative factors of other diseases. Reactive species produced in the cell during normal metabolism enter into chemical reactions with cellular biomolecules such as nucleic acids, proteins, and lipids, causing their harmful oxidative modifications. It is assumed that catalase deficiency or its defective function, which occurs for any reason, is closely related to the pathogenesis of many age-related degenerative diseases such as cancer, anemia, hypertension, Diabetes mellitus, vitiligo, Parkinson's disease, Alzheimer's disease, schizophrenia and bipolar disorder [26, 27]. Many disease phenotypes can arise from the direct or indirect degradation of the genome, transcriptome, and proteome of tissues and organs by H_2O_2. But, further research is needed to confirm this argument.

5. Peroxisomal effects on cell proliferation/differentiation, vitality, and apoptosis

Regulatory molecules involved in cell proliferation/differentiation, vitality, and apoptosis are also vulnerable to damage from oxidative attacks. Damage to molecular regulators has important consequences such as loss of proliferation control, defective cell differentiation/transformation, and apoptosis. This title was designed and shaped considering the antioxidative roles of peroxisomes against oxidative stress.

In a study with a malignant cell line, varying degrees of oxidative stress inhibited the proliferation of hepatoma cells. The number of apoptotic cells increased with the increase of oxidative stress. On the other hand, it is stated that low oxidative stress

---

[2] It is interpreted as a snake eating its own tail. It is an iconography symbolized for the eternal cyclical renewal or cycle of life, death, and rebirth, as well as the spirit of the world.

[3] Methemoglobin is a type of hemoglobin in which normal hemoglobin has ferric cation (Fe3+) instead of ferrous cation (Fe2+) in the heme group. Methemoglobinemia (MetHb) is a blood disorder in which abnormal amounts of methemoglobin are produced.

levels in all of the 4 different indices determined in hepatoma cells (such as accumulation of Con-A on the cell surface, alpha-fetoprotein, gamma-glutamyl transpeptidase, and tyrosine-alpha-ketoglutarate transaminase) result in a tendency for cell differentiation by losing some malignant features. Researchers in that study suggested the possibility that hepatoma cell growth could be inhibited, differentiation and apoptosis could be promoted, and thus reverse transformation could be initiated by tight regulation of oxidative stress [28]. In a study investigating the effect of advanced oxidation protein products (AOPP) in rat osteoblasts, it was determined by gene expression markers that proliferation and differentiation were inhibited in osteoblasts treated with AOPP [29]. It is emphasized that physiological ROS levels, which are accepted as secondary messengers, mediate numerous cellular functions in stem cells, and the need for better quantification of ROS for the correct control of stem cell fate [30]. It has been shown that miR-424, one of the miRNAs identified as key regulators of proliferation and differentiation of mesenchymal stem cells, plays a role in the regulation of bone formation *in vitro*. In this study, it was determined that the down-regulation of miR-424 under oxidative stress induced by H_2O_2 mediates bone formation [31]. It has long been known that the presence of high levels of ROS leads to impaired cell function and apoptosis. However, within the physiological range, ROS shows a wide range of variable effects. Within these effects, there are consequences ranging from preserving the strength and qualities of pure stem cells to differentiation for a particular cell group. In addition, the effects of ROS may vary in different directions according to stem cell lineage and differentiation stage. For example, the presence of ROS may be associated with decreased embryonic stem cell potency while increasing proliferation in mesenchymal stem cells and an increased likelihood of genomic instability in induced pluripotent stem cells. Again, ROS inhibits osteogenesis while increasing the differentiation of stem cells into cardiomyocytes, adipocytes, endothelial cells, keratinocytes, and neurons, and increases hypertrophic differentiation of cartilage associated with chondrocyte death [32]. Paraquat-induced oxidative stress, a known herbicide, suppressed the expression of stem cell markers including NANOG, OCT4, and TDGF1, while it increased the spontaneous expression of neuronal differentiation markers such as PAX6, NEUROD1, HOXA1, NCAM, GFRA1, and TUJ1. In addition, it has been stated that an increase in the level of intracellular ROS may trigger the exit from the stem cell state and promote neuronal differentiation of human embryonic stem cells (hESC) [33].

6. Peroxisome and immune system modulation

A healthy immune system in the individual makes effective use of ROS molecules to render many pathogens harmless. Since it is closely related to peroxisome, the known important interactions of ROS at the physiological level in the immune system should be considered.

It is known that peroxisomes, which are single membrane-enclosed organelles in the ROS turnover center, are required to inactivate bacteria by engulfing them with macrophages. Decreased peroxisome function, therefore, disrupts the rapid, temporary, large amounts of ROS production (oxidative burst) turnover needed to fight infection. This impaired response in bacterial control negatively affects the survival of the host cell and organism. Considering phagocytosis and innate immunity, a previously unknown peroxisome requirement has emerged. Dysfunction of peroxisomes in intestinal epithelial cells triggers chain events that affect cell health. This state, which increases cell death and epithelial instability, activates Tor kinase-dependent autophagy, which alters the gut microbiota, compromises immune pathways in the gut in response to infection, and affects organism survival. In addition, peroxisomes function effectively as centers that coordinate responses from metabolic, immune, and stress signaling pathways to maintain the balance in the functionality of the gut-microbe interface [34, 35]. After host cell invasion, the expression of NADPH oxidase 2 (NOX2) is induced by mycobacteria to form superoxide radicals (O^{-2}). These superoxide anions are first converted to the more toxic hydrogen peroxide (H_2O_2) via superoxide dismutase (SOD), and catalase reduces H_2O_2 to water. Contrary to what has long been assumed, accumulating evidence has shown that peroxisomes play a vital role in maintaining cellular redox balance in eukaryotic cells. Deletions in peroxisome-associated Peroxin genes impair detoxification of reactive oxygen species and post-infection peroxisome transformation, leading to the altered synthesis of transcription factors and thus various cell signaling cascades in favor of bacilli [36]. The ability of peroxisomes to modulate fatty acids explains their role as metabolic regulators in various immune functions. It has also been reported that polyamines catabolized by peroxisomes are important for T-cell clonal expansion, alternative macrophage activation, and dendritic cell modulation [37]. Peroxisomes are known to perform important roles in lipid metabolism, but recent studies focus on their essential role in modulating the immune response and inflammation [38]. At least partial peroxisome function is required for the c-Fos and NF-kB-mediated pathways in a study investigating the effects of dysfunctional peroxisomes on the infection-fighting ability of the organism. Without peroxisomes, Drosophila melanogaster has been shown to have a significantly reduced chance of survival after infection. Interestingly, it has been reported that high fatty acid concentration has an inhibitory effect on the cell's immune response [39]. The complex and dynamic interactions that exist between virus and host cells include manipulating peroxisome dynamics in the context of viral infection. Different viruses take advantage of specific peroxisome properties to counteract the host's antiviral response/promote virus particle formation and spread; to this end, viruses modulate peroxisome biogenesis and metabolism. Different studies with the same virus report opposite results, as in SARS-CoV-2 infection. For example, one study reported an increase in the number of peroxisomes, while another study pointed to peroxisome depletion. These differences are likely due to the different stages of infection analyzed. As with human cytomegalovirus (HCMV), probably many viruses interact differently with peroxisomes at different stages of their infection cycle. Viruses can promote peroxisome number depletion early in infection to inhibit host antiviral signaling, but then stimulate peroxisome metabolism and biogenesis to increase lipid metabolism and promote the formation of new virus particles [40].

7. Peroxisome and mitochondria

There are strong interactions between mitochondria and peroxisomes in terms of the production and removal of reactive molecules. Given that H_2O_2 can rapidly cross the peroxisomal and mitochondrial membrane, it is reasonable to expect that changes in peroxisomal or mitochondrial H_2O_2 metabolism will also affect other organelle functions. Whether peroxisomal H_2O_2 acts directly or indirectly on mitochondria remains to be determined. In this context, inhibition of peroxisomal catalase not only rapidly increases mitochondrial oxidative damage but also decreases the expression of

PPAR-γ co-activator (PGC) 1α, a key regulator of mitochondrial biogenesis and function. Moderate levels of mitochondrial ROS are upregulated by multiple antioxidant enzymes (nuclear respiratory factor (NRF) 2 and Forkhead box O (FOXO)) including catalase. Reliable evidence has been found that this stimulation can promote the expression of stress-sensitive transcription factors that mediate stress tolerance. Peroxisome biogenesis is dramatically induced by activation of retrograde signaling pathways in respiratory failure *S. cerevisiae* without inhibition of mitochondrial ATP synthesis. Considering that peroxisomes and mitochondria play a central role in cellular lipid metabolism and lipids have multiple roles related to bioenergetics, cellular signaling, membrane structure and function, changes in lipid metabolism of either peroxisome or mitochondria affect the function of the other organelle. A time- and inhibitory concentration-dependent pattern of pharmacological inhibition of mitochondrial β-oxidation and upregulation of peroxisomal β-oxidation in human and rat liver has been demonstrated [41]. It has also been shown that impaired mitochondrial fatty acid oxidation in skeletal muscle leads to a compensatory increase in peroxisomal fatty acid oxidation [42]. Complex peroxisomal and mitochondrial processes include common substrates (e.g., FAD, NAD^+, O_2 ve α-ketoglutarate) and metabolites (e.g., acetyl-CoA, succinate) that have the potential to directly or indirectly modulate the metabolic activities of other subcellular compartments. Many of these common substrates and metabolites can also serve as substrates or inhibitors of DNA methyltransferases, histone (de)methyltransferase/ (de)acetylases. Thus, changes in peroxisomal or mitochondrial activity are likely to affect other organelle activity through epigenetic remodeling. It has long been reported that mitochondria can communicate with the cell through the release of cytochrome c, which has a central role in apoptotic signaling. In addition, it should be well known that other death-promoting factors in the mitochondrial inner membrane space could be released into the cytosol upon induction of apoptosis. Recently, it has been shown that peroxisomes can release matrix proteins into the cytosol. Surprisingly, this release appears to be dependent on voltage-dependent anion-selective channel 2 (VDAC2), a redox-sensitive outer mitochondrial membrane protein. Loss of VDAC2 in Chinese hamster ovary cells shifts the localization of BCL2-antagonist/killer (BAK) 1, a B-cell lymphoma family member central to the mitochondrial pathway of apoptosis, from mitochondria to peroxisomes. This localization shift increases peroxisomal membrane permeability in a manner similar to that in mitochondria, resulting in the release of peroxisomal matrix proteins, including catalase, into the cytosol. It has become increasingly clear that defects in peroxisome biogenesis, peroxisomal fatty acid metabolism, or peroxisomal antioxidant capacity have a negative impact on mitochondrial function. Such mitochondrial defects can be generalized as abnormal cristae, decreased membrane potential and respiratory rates, increased ROS production, decreased fatty acid oxidation, and DNA depletion. In addition, it has been reported to induce ultrastructural and/or functional mitochondrial changes such as mass gain in various organs (e.g., brain, liver, and kidney) and cell types (e.g., skeletal and smooth muscle cells). It shows that changes in peroxisome turnover rates can affect human pathophysiology. Increased peroxisome disruption leads to worsening mitochondrial health. Here, the importance of functional peroxisomes for the maintenance of mitochondrial health is prominently emphasized. Recently, it has become increasingly clear that peroxisomes and mitochondria cooperate to fight viral infections through activation of the RIG-I-like receptors-mitochondrial antiviral signaling (RLR-MAVS) pathway. There is strong evidence to suggest that both peroxisomal and mitochondrial dysfunction may contribute to aging and age-related diseases of the organism. It is recognized that peroxisomes serve as guardians of mitochondrial health during cellular aging and age-related disease development in patients suffering from congenital peroxisomal disorders [43]. It shows that in the fungus *Podospora anserina*, the activities of peroxisomes and mitochondria in the required process for different stages of sexual development are interrelated. Peroxisomes and mitochondria share proteins that mediate the division of this fungus [44]. The family of phosphatidylinositol 5-phosphate 4-kinases (PI5P4Ks), a class of phosphoinositide kinases, phosphorylates PI-5-P to PI-4,5-P2. This phosphorylated molecule regulates peroxisomal fatty acid oxidation by mediating the transport of lipid droplets to peroxisomes, which is essential for maintaining mitochondrial metabolism [45].

8. Peroxisome and cancer

Loss of cell division control due to disruptions in the mechanism regulating cell proliferation may cause cancer, which is a devastating disease of this century. Studies focused on the role of the peroxisome in the loss of this control have pointed to peroxisome-proliferator-activated receptor (PPAR) functions. PPARs are nuclear hormone receptors that mediate the effects of fatty acids and their derivatives at the transcriptional level. Through these pathways, PPARs can regulate cell proliferation, differentiation, and survival, thereby controlling carcinogenesis in various tissues [46]. PPARs are also ligand-activated transcription factors involved in the regulation of glucose and lipid homeostasis, inflammation, proliferation, and differentiation. All of these functions draw attention to the influence of PPARs in carcinogenesis [47]. Studies on peroxisomes, which are cellular organelles that affect cancer cell growth and survival, have investigated whether the expression of peroxisome-related genes changes in more than one tumor type. The results of research gave rise to the idea that peroxisomal proteins and their metabolites may support pro-tumorigenic functions. Overexpression of several peroxisomal-associated mRNAs has been reported across a wide range of cancer types (melanoma, breast, non-small cell lung, ovarian, pancreatic, and prostate cancers). In more than 10% of all tumor types included in the analysis, progesterone-induced decidual protein (DEPP) expression was at least two-fold higher than the mean expression level. Due to its ability to indirectly induce ROS-mediated DNA damage and meet a certain threshold of pro-tumorigenic genomic instability, DEPP may be relatively abundant in multiple tumor types. Moreover, elevated DEPP can activate autophagy, an aberrant process in tumorigenesis, which is particularly upregulated in treatment-resistant cancers. Similarly, PEX16 protein appears to be relatively elevated in tumors compared to normal tissue [48–50]. In different studies investigating the role of peroxisomes in spontaneously formed tumor tissue, it was observed that peroxisomal function is decreased in neoplastic tissue such as colon carcinoma when compared to unaffected tissue. Publications are reporting the amount of peroxisomal protein (catalase, ABCD3, ACOX1, PXMP2) or enzyme (catalase, D-amino acid oxidase, polyamine oxidase, peroxisomal β-oxidation) activities are decreased in breast and hepatocellular carcinomas similarly in colon tumor tissue. A high α-methylacyl-CoA racemase (AMACR) level has been suggested as a reliable prostate cancer tumor marker, as the expression of peroxisomal AMACR was found to be quite high in tissue from prostate carcinoma compared to benign prostate tissue. Not only prostate, but high AMACR expression has also been reported from stomach, colon, breast, kidney, and hepatocellular carcinoma. In the same review, the peroxisomal membrane protein PMP24/PXMP4 has been associated with prostate cancer development. Here, PMP24 is a member of the TIM17 family of membrane proteins. Peroxisomes in glioblastomas have been investigated for tumor-grade progression. Increased immunocytochemistry staining was observed for peroxisome increase associated with progressive tumor grade. This finding was confirmed by the detection of Pex14, PMP70, ACOX1, and 3-ketothiolase proteins, which indicate increased organelle number, namely peroxisome proliferation, using immunoblotting [51]. It was observed that peroxisomal Lon peptidase (LonP2) expression increased in cervical cancer tissue. Basically, LonP2 functions as a combined chaperone/protease by refolding or cleaving degraded peroxisomal proteins. Downregulation of LonP2 in HeLa and SiHA tumor cell lines reduced oxidative stress and inhibited cervical cancer cell proliferation and migration [52, 53]. An unexpected function of peroxisomes in the control of cell division that may be associated with tumor progression is discussed. It has been demonstrated that the correct positioning of peroxisomes during mitosis is required for asymmetric cell division in skin epithelial cells. RNAi-mediated knock-down of Pex11β and Pex14 induced mitotic delay in targeted cells and resulted in an imbalance in growth and differentiation in basal and supra-basal skin cells, together with a reduction in terminal differentiation markers in tissue. The detected mitotic disorder was not associated with an impairment in peroxisomal functions but was found to result from the misplacement of peroxisomes during spindle formation [54].

A study reporting that peroxisomes are indispensable for the survival of liver cancer cells reported that tumor growth was significantly reduced by RNAi silencing of Pex2 in hepatocellular carcinoma xenografts [55]. Consistent with this data, decreased catalase activities were also found in kidney tumors [56]. Peroxisomal protein levels or enzyme activities are greatly reduced in some types of cancer (such as renal, breast, hepatocellular, and colon cancer). Here, hypoxia-inducible transcription factor (HIF-2α) has been shown to promote peroxisome degradation. A decrease in peroxisome abundance was observed in renal carcinoma cells with high HIF-2α levels. These studies report a reduction in peroxisomal activity in some tumor types, while other reports suggest that the metabolic activity of the peroxisome promotes tumor growth. The tumor-promoting or tumor-suppressing function of the peroxisome probably depends on the tumor type in the particular microenvironment. However, peroxisomal genes may be the target of a potential anticancer therapeutic strategy due to their functional role in tumorigenesis. One of the most prominent candidates as a target is alkylglycerone phosphate synthase (AGPS), which is involved in peroxisomal ether lipid biosynthesis. Inactivation of AGPS reduced the levels of ether lipids, including plasmalogens, in breast cancer and melanoma cells, inhibiting their tumorigenicity *in vitro* and *in vivo*. In addition, loss of AGPS decreased the invasive capacity by down-regulating ether lipid expression in glioma and liver cancer cells. Therefore, peroxisomes, which are the center of the antioxidant system, have the potential to be an anticancer target in combination with chemotherapy to generate oxidative stress. Considering many data, it is considered that cancer cells are prone to ROS-induced apoptosis, and factors that impair peroxisome integrity/function may increase the success of cancer therapy. For example, PEX3 degradation predisposes lymphoma cells to ROS-induced apoptosis. Similarly, the loss of proteins such as PEX2 and PEX5 involved in peroxisome protein import stimulates apoptosis in hepatocellular carcinoma (HCC) [57]. In a new study to determine the prognostic value of the peroxisomal pathway in colorectal cancer (CRC), the combined evaluation of T-cell immunoglobulin and mucin domain 3 (TIM3) expression and genes involved in the peroxisome pathway or Fatty acid alpha oxidation (FAAO). Data from here can be used for diagnosis and can be helpful for personalized treatment [58].

In summary, the roles of peroxisomes with pleiotropic effects should be evaluated separately for healthy and cancer cell types. ROS molecules, which are dangerous for healthy cells in the survival process, are also dangerous for cancer cells and their apoptotic effect is dominant. ROS, which is an enemy for healthy cells, is also an important enemy for cancer cells and thus is the enemy of our enemy. This situation, which offers an important perspective for the homeostasis of the organism, reminds us of the proverb "Amicus meus, inimicus inimici mei" (Enemy of my enemy is my friend) that was first mentioned in the work on state administration called "Arthashastra" written in Sanskrit in the 4th century BC.

9. Stimulation or inhibition of peroxisomes' abundance

Peroxisome abundance is regulated dynamically according to the general conditions of the cell. This regulation is sometimes stimulation and sometimes inhibition. Cell requirements and complex interactions with other cell components determine the direction of this regulation.

Consensus has been reached that the main mode of peroxisome proliferation in yeast is fission. Also, in mammals, fission is most likely the main mode of peroxisome proliferation, although *de novo* synthesis can also occur. *De novo* peroxisome synthesis from the ER was investigated by the reintroduction of the PEX3 gene into PEX3 mutant or peroxisome-deficient yeasts [59].

After treatment of fibroblasts with the peroxisome biogenesis disorder (PBDs) phenotype with 4-phenylbutyrate, an approximately two-fold increase in peroxisome number, again an increase in transcription of the adrenoleukodystrophy-related gene and PEX11alpha was found. In addition, there was an increase in very-long-chain fatty acid beta-oxidation and plasmalogen concentrations and hence a decrease in very-long-chain fatty acid concentrations in fibroblasts of patients with PBD.

In the process of pexophagy, a type of macroautophagy that selectively degrades peroxisomes, double-membraned autophagosomes surround the peroxisomes, fusing them with lysosomes for degradation. Yeasts (*S. cerevisiae, P. pastoris,* and *Hansenula polymorpha*) can use different carbon sources to obtain energy, so they rapidly increase biogenesis-related peroxisome abundance and the formation of giant peroxisome clusters when grown in environments with oleic acid, methanol, oleate, or amines based on peroxisome metabolism. Conversely, the shift of yeast from peroxisome-dependent carbon sources, such as lipids, to peroxisome-independent carbon sources, such as glucose, triggers their abrupt degradation by pexophagy. Similarly, treating rodents with peroxisome-enhancing stimuli (phthalate esters and hypolipidemic drugs) rapidly increases peroxisome abundance, again triggering large-scale pexophagy upon removal of the stimulus. Pexophagy plays a critical role in peroxisome quality control, as peroxisomes have a half-life of 1.5–2 days in mammals. Failure to import matrix proteins (Pex1p, Pex6p, and Pex15p) identifies peroxisomes for quality control pexophagy. By the accumulation of ubiquitinated Pex5p on the cytosolic surface of the peroxisomal membrane, the pexophagy receptor appears to induce autophagy. However, it has been suggested that other signals, such as phosphorylation of some of the autophagy-related (Atg) proteins in yeast, could be the pexophagy receptor. Overexpression of C-terminal EGFP-tagged PEX5 in transformed mouse embryonic fibroblasts (MEFs) stimulates accumulation and monoubiquitination of PEX5-EGFP in the peroxisome, resulting in pexophagy. During oxidative stress, PEX5 is phosphorylated via mutant ataxia-telangiectasia (ATM) proteins, making it possible to add ubiquitin and subsequently target it by autophagosomes. Yeast pexophagy is regulated via glucose-sensitive and mitogen-activated protein kinase (MAPK) cascades, but the precise mechanisms of these cascades are not fully understood. Although pexophagy is generally viewed as a similar pathway in yeast and mammals, there are distinct differences in the way the pexophagy pathway is governed. Mammalian pexophagy receptors NBR1 and p62 selectively target ubiquitinated peroxisomes for pexophagy, while Atg30p or Atg36p on yeast phospho-activated pexophagy receptors interact with Pex3p and Atg37p on peroxisomes to mediate pexophagy. Mammals allow differentiation between peroxisomes based on the state of ubiquitin accumulation, while it is not clear whether this type of mechanism exists in yeast [60].

10. Conclusion and projection

The peroxisome, which is at the center of the lipid metabolism and antioxidant defense system in the cell, is a versatile organelle surrounded by a single membrane. Numerous components work in concert for a flawless and sustainable peroxisome function. Most of these components are Peroxin (PEX) group molecules, and they have signal, transport, and receptor functions in protein import. Some components have structural functions in the peroxisome, and some have enzymatic functions. Syndromes due to peroxisome defects usually represent a heterogeneous group of congenital diseases. Interestingly, mutations in some components of the peroxisome import pathways are not only associated with yeast but also with serious human diseases including peroxisome-related disorders. In some such diseases, only one peroxisomal enzyme is deficient. However, in some other diseases that result from defects in peroxisome function, multiple peroxisomal enzymes cannot be transferred to peroxisomes or are caused by deficiencies in the PTS1 or PTS2 pathways responsible for peroxisome protein transport.

Despite the large body of data available, we are still far from the goal of a full understanding of the peroxisome nature, which is surprisingly plastic and pleiotropic. Each new experimental study illuminates the peroxisome mystery and opens new horizons in terms of interaction between organelles and their effects on cell homeostasis. With the expansion of the peroxisome knowledge pool, we will have a better understanding of the organelle's function and subcellular interaction dynamics, and we will have advanced instruments for new revolutionary treatment strategies.

Conflict of interest

The author declared no financial or commercial conflict of interest.

Author details

Hasan Basri İla
Faculty of Science and Letter, Department of Biology, Çukurova University, Adana, Turkey

*Address all correspondence to: ila@cu.edu.tr; hasanbasriila@gmail.com

References

[1] Kump LR. The rise of atmospheric oxygen. Nature. 2008;**451**(7176):277-278

[2] Farquhar J, Bao H, Thiemens M. Atmospheric influence of Earth's earliest sulfur cycle. Science. 2000;**289**(5480):756-759

[3] Berner RA, Vandenbrooks JM, Ward PD. Evolution. Oxygen and evolution. Science. 2007;**316**(5824): 557-558

[4] Lee S et al. Mitochondrial H2O2 generated from electron transport chain complex I stimulates muscle differentiation. Cell Research. 2011;**21**(5):817-834

[5] Bao L et al. Mitochondria are the source of hydrogen peroxide for dynamic brain-cell signaling. The Journal of Neuroscience. 2009;**29**(28):9002-9010

[6] Bonekamp NA et al. Reactive oxygen species and peroxisomes: Struggling for balance. BioFactors. 2009;**35**(4):346-355

[7] Wanders RJ, Waterham HR. Biochemistry of mammalian peroxisomes revisited. Annual Review of Biochemistry. 2006;**75**:295-332

[8] Brites P, Waterham HR, Wanders RJ. Functions and biosynthesis of plasmalogens in health and disease. Biochimica et Biophysica Acta. 2004;**1636**(2-3):219-231

[9] Antonenkov VD. Dehydrogenases of the pentose phosphate pathway in rat liver peroxisomes. European Journal of Biochemistry. 1989;**183**(1):75-82

[10] Cooper JA. Peroxisome [online]. 2020. Available from: https://www.britannica.com/science/peroxisome

[11] Charles KN et al. Functional peroxisomes are essential for efficient cholesterol sensing and synthesis. Frontiers in Cell and Development Biology. 2020;**8**:560266

[12] Cooper GM. The Cell - a Molecular Approach. 2nd ed. Sunderland (MA): Sinauer Associates; 2000

[13] Jansen RLM et al. Comparative genomics of peroxisome biogenesis proteins: Making sense of the PEX proteins. Frontiers in Cell and Development Biology. 2021;**9**:654163

[14] Zarsky V, Tachezy J. Evolutionary loss of peroxisomes - not limited to parasites. Biology Direct. 2015;**10**

[15] Farre JC et al. Peroxisome biogenesis, membrane contact sites, and quality control. EMBO Reports. 2019;**20**(1)

[16] Schluter A et al. PeroxisomeDB 2.0: An integrative view of the global peroxisomal metabolome. Nucleic Acids Research. 2010;**38**(Database issue:D800-D805

[17] Gould SJ et al. A conserved tripeptide sorts proteins to peroxisomes. The Journal of Cell Biology. 1989;**108**(5):1657-1664

[18] Petriv OI et al. A new definition for the consensus sequence of the peroxisome targeting signal type 2. Journal of Molecular Biology. 2004;**341**(1):119-134

[19] Chen BH et al. Hsc70/Stub1 promotes the removal of individual oxidatively stressed peroxisomes. Nature Communications. 2020;**11**(1)

[20] Fransen M. Peroxisome dynamics: Molecular players, mechanisms, and

(Dys)functions. ISRN Cell Biology. 2012;**2012**:714192

[21] Kwak MS et al. Reactive oxygen species induce Cys106-mediated anti-parallel HMGB1 dimerization that protects against DNA damage. Redox Biology. 2021;**40**:101858

[22] Zhang J et al. ROS and ROS-mediated cellular signaling. Oxidative Medicine and Cellular Longevity. 2016;**2016**:4350965

[23] Dansen TB, Wirtz KW. The peroxisome in oxidative stress. IUBMB Life. 2001;**51**(4):223-230

[24] Okumoto K et al. The peroxisome counteracts oxidative stresses by suppressing catalase import via Pex14 phosphorylation. eLife. 2020;**9**

[25] Sandalio LM et al. Peroxisomes as redox-signaling nodes in intracellular communication and stress responses. Plant Physiology. 2021;**186**(1):22-35

[26] Góth L, Nagy T. Inherited catalase deficiency: Is it benign or a factor in various age related disorders? Mutation Research/Reviews in Mutation Research. 2013;**753**(2):147-154

[27] Nandi A et al. Role of catalase in oxidative stress- and age-associated degenerative diseases. Oxidative Medicine and Cellular Longevity. 2019;**2019**:9613090

[28] Ren JG, Li JF, Zheng RL. Effects of oxidative stress on the proliferation, differentiation and apoptosis in the human hepatoma cells. Shi Yan Sheng Wu Xue Bao. 1998;**31**(3):273-282

[29] Sun N et al. Effect of advanced oxidation protein products on the proliferation and osteogenic differentiation of rat mesenchymal stem cells. International Journal of Molecular Medicine. 2013;**32**(2):485-491

[30] Sart S, Song LQ, Li Y. Controlling redox status for stem cell survival, expansion, and differentiation. Oxidative Medicine and Cellular Longevity. 2015;**2015**

[31] Li L et al. FOXO1-suppressed miR-424 regulates the proliferation and osteogenic differentiation of MSCs by targeting FGF2 under oxidative stress. Scientific Reports. 2017;**7**(1):42331

[32] Nugud A, Sandeep D, El-Serafi AT. Two faces of the coin: Minireview for dissecting the role of reactive oxygen species in stem cell potency and lineage commitment. Journal of Advanced Research. 2018;**14**:73-79

[33] Hu QD et al. Oxidative stress promotes exit from the stem cell state and spontaneous neuronal differentiation. Oncotarget. 2018;**9**(3):4223-4238

[34] Di Cara F et al. Dysfunctional peroxisomes compromise gut structure and host defense by increased cell death and tor-dependent autophagy. Molecular Biology of the Cell. 2018;**29**(22):2766-2783

[35] Di Cara F et al. Peroxisome-mediated metabolism is required for immune response to microbial infection. Immunity. 2017;**47**(1):93-106

[36] Ganguli G, Mukherjee U, Sonawane A. Peroxisomes and oxidative stress: Their implications in the modulation of cellular immunity during mycobacterial infection. Frontiers in Microbiology. 2019;**10**

[37] Di Cara F. Peroxisomes in host defense. PLoS Pathogens. 2020;**16**(7):e1008636

[38] Lucaccioni L et al. Overwhelming sepsis in a neonate affected by Zellweger syndrome due to a compound heterozygosis in PEX 6 gene: A case report. BMC Medical Genetics. 2020;**21**(1)

[39] Sara M. PEX1 and PEX7 genes are necessary for completion of immune pathways and survival post-infection. STEM Fellowship Journal. 2020;**6**(1):5-11

[40] Ferreira AR et al. Emerging roles of peroxisomes in viral infections. Trends in Cell Biology. 2022;**32**(2):124-139

[41] Vickers AEM, Bentley P, Fisher RL. Consequences of mitochondrial injury induced by pharmaceutical fatty acid oxidation inhibitors is characterized in human and rat liver slices. Toxicology In Vitro. 2006;**20**(7):1173-1182

[42] Wicks SE et al. Impaired mitochondrial fat oxidation induces adaptive remodeling of muscle metabolism. Proceedings of the National Academy of Sciences of the United States of America. 2015;**112**(25):E3300-E3309

[43] Fransen M, Lismont C, Walton P. The peroxisome-mitochondria connection: How and why? International Journal of Molecular Sciences. 2017;**18**(6)

[44] Navarro-Espindola R et al. Distinct contributions of the peroxisome-mitochondria fission machinery during sexual development of the fungus Podospora anserina. Frontiers in Microbiology. 2020;**11**

[45] Ravi A et al. PI5P4Ks drive metabolic homeostasis through peroxisome-mitochondria interplay. Developmental Cell. 2021;**56**(11):1661

[46] Michalik L, Desvergne B, Wahli W. Peroxisome-proliferator-activated receptors and cancers: Complex stories. Nature Reviews. Cancer. 2004;**4**(1):61-70

[47] Peters JM, Shah YM, Gonzalez FJ. The role of peroxisome proliferator-activated receptors in carcinogenesis and chemoprevention. Nature Reviews. Cancer. 2012;**12**(3):181-195

[48] Dahabieh MS et al. Peroxisomes and cancer: The role of a metabolic specialist in a disease of aberrant metabolism. Biochimica Et Biophysica Acta. Reviews on Cancer. 2018;**1870**(1):103-121

[49] Del Paggio JC. Immunotherapy: Cancer immunotherapy and the value of cure. Nature Reviews. Clinical Oncology. 2018;**15**(5):268-270

[50] Salcher S et al. C10ORF10/DEPP-mediated ROS accumulation is a critical modulator of FOXO3-induced autophagy. Molecular Cancer. 2017;**16**(1):95

[51] Islinger M et al. The peroxisome: An update on mysteries 2.0. Histochemistry and Cell Biology. 2018;**150**(5):443-471

[52] Bartoszewska M et al. Peroxisomal Proteostasis involves a Lon family protein that functions as protease and chaperone. Journal of Biological Chemistry. 2012;**287**(33):27380-27395

[53] Wu WR et al. Lon peptidase 2, Peroxisomal (LONP2) contributes to cervical carcinogenesis via oxidative stress. Medical Science Monitor. 2018;**24**:1310-1320

[54] Asare A, Levorse J, Fuchs E. Coupling organelle inheritance with mitosis to balance growth and differentiation. Science. 2017;**355**(6324)

[55] Cai M et al. Disruption of peroxisome function leads to metabolic stress, mTOR inhibition, and lethality in liver cancer cells. Cancer Letters. 2018;**421**:82-93

[56] Pljesa-Ercegovac M et al. Altered antioxidant capacity in human renal cell carcinoma: Role of glutathione associated enzymes. Urologic Oncology. 2008;**26**(2):175-181

[57] Kim JA. Peroxisome metabolism in cancer. Cell. 2020;**9**(7)

[58] Yin J et al. Identification of an at-risk subpopulation with high immune infiltration based on the peroxisome pathway and TIM3 in colorectal cancer. BMC Cancer. 2022;**22**(1):44

[59] Veenhuis M, van der Klei IJ. A critical reflection on the principles of peroxisome formation in yeast. Frontiers in Physiology. 2014;**5**:110

[60] Germain K, Kim PK. Pexophagy: A model for selective autophagy. International Journal of Molecular Sciences. 2020;**21**(2)

Section 2

Peroxisomal Diseases and Treatment Strategies

Chapter 3

Apoptosis-Related Diseases and Peroxisomes

Meimei Wang, Yakun Liu, Ni Chen, Juan Wang and Ye Zhao

Abstract

Apoptosis is a highly regulated cell death program that can be mediated by death receptors in the plasma membrane, as well as the mitochondria and the endoplasmic reticulum. Apoptosis plays a key role in the pathogenesis of a variety of human diseases. Peroxisomes are membrane-bound organelles occurring in the cytoplasm of eukaryotic cells. Peroxisomes engage in a functional interplay with mitochondria. They cooperate with each other to maintain the balance of reactive oxygen species homeostasis in cells. Given the key role of mitochondria in the regulation of apoptosis, there could also be an important relationship between peroxisomes and the apoptotic process. Peroxisome dysfunction severely affects mitochondrial metabolism, cellular morphological stability, and biosynthesis, and thus contributes directly or indirectly to a number of apoptosis-related diseases. This chapter provides an overview of the concept, characteristics, inducing factors, and molecular mechanisms of apoptosis, as well as evidence for apoptosis in cancer, cardiovascular diseases, and neurodegenerative disorders, and discusses the important role of the peroxisome in the apoptosis-associated diseases.

Keywords: apoptosis, mitochondria, peroxisome, ROShomeostasis, Cancer, cardiovascular diseases, neurodegenerative diseases

1. Introduction

Death is the final fate of cells and organisms and is a normal biological phenomenon in the living world. Cell death plays a crucial role in the development of plants and animals in nature and in maintaining ecological balance [1]. For example, in the developing vertebrate nervous system, as many as half or more of the nerve cells usually die soon after they are formed. In a healthy adult human, billions of cells die every hour in the bone marrow and intestines. So much cell death seems very wasteful, especially when the vast majority of cells are perfectly healthy at the time of suicide.

In general, cell death can be divided into two types: programmed cell death (PCD) and accidental cell death (necrosis) [2]. The former is a controlled process of intracellular death program, also vividly referred to as cellular suicide. The latter is caused by external factors (i.e., injury, infection, etc.). The study of PCD (especially apoptosis) processes has led to a better understanding of the pathogenesis of certain diseases. The 2002 Nobel Prize in Physiology and Medicine was awarded to Britons Sydney Brenner, Jone E. Sulston, and H Robert Horvitz for their discovery of how genes regulate organ growth and programmed cell suicide processes, using the nematode *Caenorhabditis elegans* (*C. elegans*) as an animal model.

A coordinated balance between cell proliferation and apoptosis is crucial for normal development and tissue homeostasis. Once this balance is permanently disrupted, normal cells may be transformed into mutant cells whose clonal survival and uncontrolled proliferation may lead to the development of tumors and various other diseases.

2. Apoptosis

2.1 The concept, characteristics, and inducing factors

2.1.1 Key concepts

Apoptosis is the process of cellular suicide by activating an intracellular death program or by the orderly breakdown of cells from within. The term was first introduced by Kerr J. F. R. in the 1970s and was not accepted by the general public until the 1990s.

Although apoptosis is only one form of Programmed cell death (PCD), it is by far the most common and well-understood form, and, confusingly, biologists often use the terms PCD and apoptosis interchangeably [3].

For a multicellular organism, a highly organized community, cell numbers are tightly regulated not only by controlling the rate of cell division but also by controlling the rate of cell death. Thus, apoptosis is important not only for tissue remodeling and elimination of transitional organs during the development of an organism, but also for the clearance of cellular senescence inactive metabolic organs, such as blood cells and epithelial cells in the digestive system, and cells with damaged or mutated DNA [4–6]. In a nutshell, apoptosis is an essential mechanism complementary to proliferation to ensure homeostasis in all tissues.

Unlike apoptosis, necrosis is a form of cell injury that leads to the premature death of cells in living tissues due to autolysis, usually caused by stronger external factors such as infection, toxins, or trauma, ultimately resulting in the unregulated of cellular components, always harmful and potentially fatal to the organism [7, 8]. Necrosis usually causes a local inflammatory response. The reason for this is that when nearby macrophages engulf these necrotic cells, they may release microorganisms that destroy the surrounding tissue causing collateral damage and inhibiting the healing process.

Typically, cell death due to necrosis does not follow the apoptotic signaling transduction pathway, but rather various receptors are activated, leading to loss of cell membrane integrity and uncontrolled release of cell death products into the extracellular space. In contrast, apoptosis is a naturally occurring programmed and targeted cause of cell death and usually provides beneficial effects to the organism. A brief comparison of them can be summarized as follows (**Figure 1**).

2.1.2 Characteristics

As mentioned above, necrosis is a form of traumatic cell death caused by acute cellular injury. In contrast, apoptosis is a process of active cellular suicide. Multicellular organisms eliminate mutated, damaged, or unwanted cells by this type of active suicide. Apoptosis plays an important role in tissue sculpting during embryonic development and in the maintenance of tissue homeostasis throughout life [6].

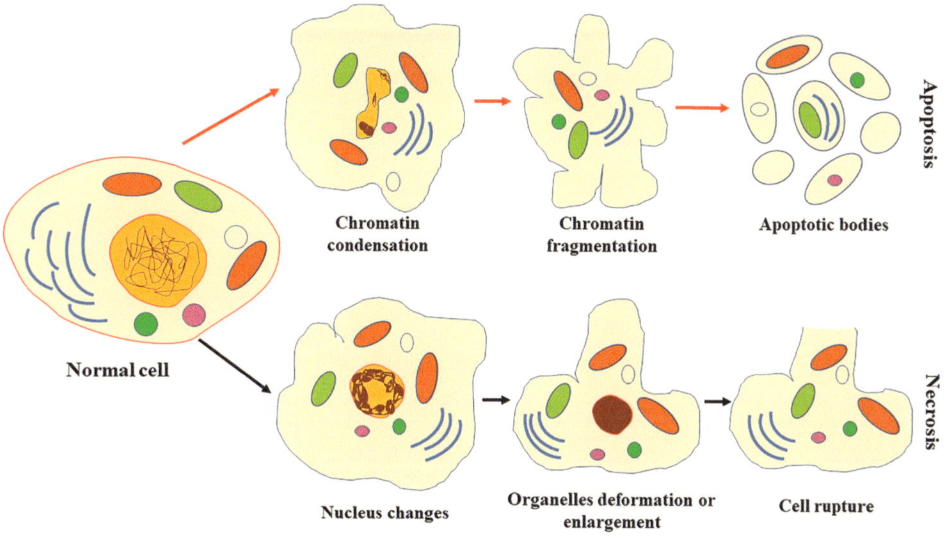

Figure 1.
Structural change of cells undergoing necrosis and apoptosis.

The process has distinct morphological features, including cell rounding and contraction, blebbing and PS externalization of the plasma membrane, cytoplasmic vacuolization including endoplasmic reticulum expansion and cisternae swelling to form vesicles and vacuoles, nuclear condensation, border aggregation or fragmentation, chromatin compaction, pyknosis, and ultimately fragmentation between nucleosomes by endonucleases, resulting in regular DNA degradation and inhibition of protein translation, and ultimately to the eventual rupture of the cell into small spheres surrounded by membranes called apoptotic bodies, which contain "packed" cell contents with an electron cloud density similar to chromatin; and a sub-G1 curve preceding the G1 phase peak is observed in cytometric histogram [9]. Apoptotic bodies can be recognized and digested by phagocytosis of neighboring macrophages through the presence of phosphatidylserine (PS) on their surface [10]. In this way, the apoptotic cells can be rapidly removed by tissue phagocytes through phagocytosis, without releasing harmful substances that can initiate inflammation, which can cause a significant amount of tissue damage. Because apoptotic cells are always rapidly eaten and digested, dead cells are usually rarely seen, even when large numbers of cells die from apoptosis. This may be the reason why biologists once ignored the phenomenon of apoptosis and may still underestimate its extent.

Abnormal apoptosis contributes to many important diseases, including cancer, autoimmune diseases, diabetes, and neurodegenerative diseases. Various types of cellular stress, such as DNA damage or growth factor deprivation, can trigger apoptosis through intrinsic or extrinsic pathways.

2.1.3 Inducing factors

Apoptosis can be triggered by both internal stimuli, such as abnormalities in DNA, and external stimuli, such as certain cytokines from different pathways, respectively [11]; or it can be induced by physiological or pathological factors.

Specifically, physiological triggers can include the following two aspects [12]:
(1) Direct action of certain hormones and cytokines: for example, glucocorticoids

are typical signals of apoptosis in lymphocytes; thyroxine plays an important role in the apoptotic degeneration of tadpoles' tails; TNF can induce apoptosis in a variety of cells. (2) Indirect effects of certain hormones and cytokines: for example, testosterone deficiency caused by testicular dysplasia can lead to apoptosis of prostate epithelial cells. Inadequate secretion of adrenocorticotropic hormone by the pituitary gland can promote apoptosis of adrenocortical cells, etc.

While pathological triggers usually include the following two aspects: (1) It is generally believed that apoptosis can be induced by many factors that can cause damage to cells, such as stress, radiation, chemical toxins, viral infections, and chemotherapeutic drugs, and even malnutrition and excessive functional complexes can induce apoptosis. (2) Some factors such as various chemical carcinogens and certain viruses (e.g., EBV) inhibit apoptosis. Therefore, it is thought that the ability to induce cells may be related to the type, intensity, and duration of the harmful factors.

2.2 Molecular mechanisms of apoptosis (Signaling pathways and related enzymes)

2.2.1 Apoptotic Signaling pathways

The initiation of apoptosis is tightly regulated by different signaling pathways. The best-understood two are the intrinsic pathway (also known as the mitochondrial pathway) and the extrinsic pathway (also known as the death receptor pathway). The mitochondrial pathway is generally activated by intracellular signals and depends on proteins released from the intermembrane space between the mitochondrial bilayers. The death receptor pathway is activated by extracellular ligands, and the activated extracellular ligands bind to their specific death receptors on the cell surface, inducing the formation of death-inducing signaling complexes (DISC) [13, 14]. Here, we will discuss the extrinsic and intrinsic pathways separately. However, it should be noted that there is crosstalk between these pathways and that extracellular apoptotic signaling can also lead to activation of the intrinsic pathway.

2.2.1.1 The extrinsic pathway of apoptosis

The extrinsic death pathway triggers receptor-mediated apoptosis. Its major components include pro-apoptotic ligands, receptors that recognize/bind ligands, and adaptor proteins that bind to the cytoplasmic face of the receptor. In addition, the pathway recruits other molecules, including cysteine-specific proteases (caspases), the initiator of the death process, and the executors, to execute the apoptotic process [15]. For example, TNF is a common pro-apoptotic ligand and TNFR1 on the cell membrane is the receptor. When TNF binds to TNFR1, the activated receptor binds to two different cytoplasmic adaptor proteins (tumor necrosis factor-related death domain protein, TRADD, and fas-associating protein with death domain, FADD) and procaspase-8, forming a multi-protein complex on the inner surface of the plasma membrane, containing an 80 amino acid death structure domain through which a death-inducing signaling complex (DISC). The cytoplasmic structural domains of the TNF receptor, FADD, and TRADD interact through homologous regions called death structural domains present in each protein [16]. Procaspase-8 and FADD interact through homologous regions called death effector domains. Procaspase 8 in DISC is activated and active caspase 8 is released into the cytoplasm, where it cleaves and activates effector caspases (e.g., procaspase 3), triggering a caspase cascade that further cleaves a number of death substrates, including BID and cytoskeletal proteins, if glued, leading to apoptosis (**Figure 2**). Notably, inhibitors of apoptosis (IAPs) can inactivate caspases by specifically binding to their active sites. Caspase activator (SMAC)/Diablo and its functional homologs in flies, including Grim, Reaper, and Hid, can in turn target binding and degrade IAPs [17].

In addition, it should be noted that the interaction between TNF and TNFR1 may also activate other signaling pathways and allow cell survival rather than self-destruction.

2.2.1.2 The intrinsic pathway of apoptosis

In general, internal stimuli such as irreparable genetic damage, hypoxia (lack of oxygen), very high concentrations of cytosolic Ca^{2+}, viral infection, or severe oxidative stress (i.e., production of large amounts of damaging free radicals) and cytotoxic drug treatment trigger apoptosis via the intrinsic pathway.

The intrinsic death pathway, i.e., the mitochondrial-received apoptotic pathway, is a death receptor non-dependent apoptotic pathway [18]. This pathway is activated by the release of cytochrome C (Cyto C) from mitochondria in response to various stresses and developmental death cues. The process specifically involves multiple steps as follows: apoptotic signals (various types of cellular stress), lead to the insertion of pro-apoptotic members of the Bcl-2 family of proteins (e.g., Bax), into the outer mitochondrial membrane, forming pores that mediate the release of Cyto C from the mitochondrial intermembrane space into the cell membrane. Once in the cell membrane, Cyto C molecules bind to Apaf-1 (a homolog of mammalian CED4) and further recruit procaspase-9 to form a complex of multiple subunits called the apoptosome. Then procaspase-9 is activated to become active caspase-9. Then the caspase-9 molecule cleaves and activates the downstream executor caspase (Caspase-3, 6,7) to carry out the apoptotic process (**Figure 2**) [19].

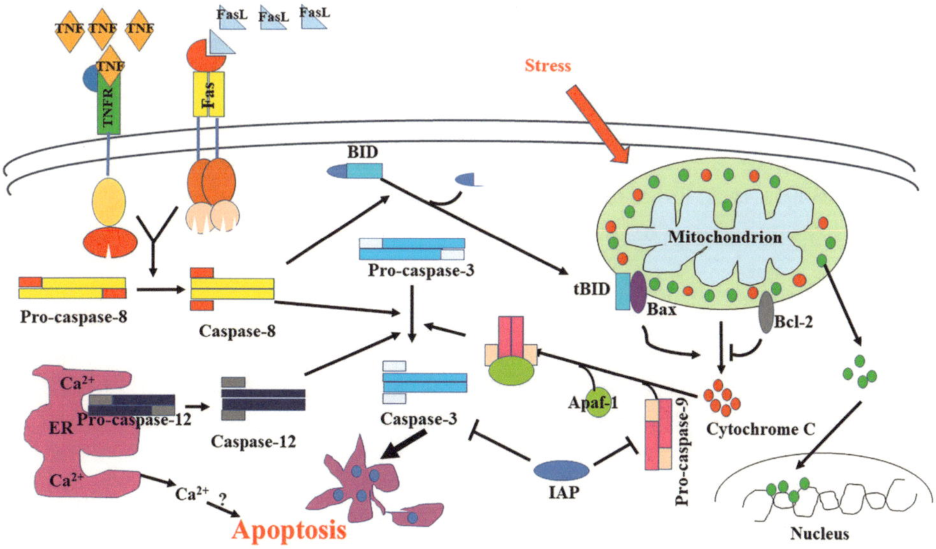

Figure 2.
Schematic diagram of apoptotic signaling.

Bcl-2, the mammalian homolog of Ced-9, prevents apoptosis by inhibiting the release of CytoC from mitochondria [20]. IAPs, second mitochondrial activators of caspases (Smac), endonuclease G (Endo G), and AIF also have important roles in the apoptotic process [21]. Notably, Endo G and AIF are specifically activated by apoptotic stimuli and are able to induce ribosomal breakage of DNA independently of caspases. Endo G is a mitochondria-specific nuclease that translocates to the nucleus and cleaves chromatin DNA during apoptosis. AIF is a flavin adenine dinucleotide-containing, NADH-dependent oxidoreductase that resides in the mitochondrial intermembrane space, and its specific enzymatic activity remains unknown. In the presence of apoptosis, AIF undergoes proteolysis and translocates to the nucleus, where it triggers chromatin condensation and massive DNA degradation in a caspase-independent manner.

2.2.1.3 ER-dependent apoptotic pathways

Previously, it was thought that the only apoptotic pathways were the mitochondrial pathway and the death receptor signaling pathway. Now, an increasing number of studies have shown that the endoplasmic reticulum (ER) also senses and transmits apoptotic signals [22, 23]. The sustained action of various apoptosis-inducing factors may induce a complex unfolded protein response (UPR) by interfering with the correct protein folding process. The UPR response causes endoplasmic reticulum stress, leading to cellular apoptosis due to the accumulation of intracellular misfolded proteins. ER, in addition to being the site of protein folding, it is also the main intracellular Ca^{2+} reservoir. Disturbing intracellular Ca^{2+} homeostasis can also induce the typical ER stress response. Interestingly, the localization of Bcl-2 family proteins (including Bcl-1, Bax, Bak, *et al.*) in the ER affects the cellular Ca^{2+} homeostasis. Overexpression of Bcl-2 or deficiency of Bax and Bak decreases the Ca^{2+} concentration in the ER and protects cells from apoptotic stimuli that trigger cell death by inducing Ca^{2+} influx from the ER to the cell membrane (**Figure 2**).

It has been suggested that procaspase-12 is a proximal effector of apoptosis associated with the ER. Recent studies have found that although caspase-12 is processed and activated in ER stress-induced apoptosis in mouse cells, the enzyme is not absolutely necessary for this process. On the other hand, cells lacking caspase-8 or caspase-9 were highly resistant to ER stress-induced apoptosis. One of the mechanisms that could explain caspase-8 activation in the ER involves the recent discovery of an ER-resident potential apoptosis initiator, named neurotrophic receptor-like death domain protein (NRADD). This protein has a transmembrane and cytoplasmic region that is highly homologous to the death receptor. Induction of apoptosis by NRADD is dependent on caspase-8 activation but does not require the mitochondrial component of the death program.

In addition to propagating death-inducing stress signals, ER contributes to apoptosis initiated by cell surface death receptors and to pathways resulting from DNA damage. Modulation of ER calcium stores can sensitize mitochondria to direct pro-apoptotic stimuli and promote activation of cytoplasmic death pathways.

In short, the extrinsic (receptor-mediated), intrinsic (mitochondria-mediated), and endoplasmic reticulum stress-mediated apoptotic pathways ultimately converge by activating the same caspases, which cleave the same cellular targets. Apoptosis-inducing factors can be involved in diseases by activating apoptotic pathways that affect the rate of apoptosis, and may predominantly involve the first two pathways or all three of these pathways.

2.2.2 Apoptosis-related enzymes

The mechanism by which apoptosis occurs is highly conserved in all animal cells. It is dependent on a family of proteins called caspases (c for cysteine and asp for aspartic acid). This family of proteins has many members and generally exists as inactive precursors (procaspases). Procaspases are generally activated by the catalytic cleavage of other (already active) caspases, forming an amplified network of protein cascades. The activation process of procaspases involves the formation of a heterodimer by cleavage and the combination of two dimers to form an active tetramer. During apoptosis, those responsible for initiation are known as initiator caspases; those responsible for cleavage of specific target proteins (e.g. nuclear lamina proteins, DNA degradation enzymes, cytoskeletal proteins, and cell–cell adhesion proteins) are the executor caspases.

Apoptotic mechanisms are present throughout the initial to final stages of animal development. Only the process requires a trigger to be activated for its occurrence. So, how is the first member of the caspase cascade reaction described above initiated? Initiator procaspases usually contain a caspase recruitment domain (CARD). This structural domain can assemble into an activation complex with an adaptor protein when the cell receives an apoptotic signal. The formation of this complex means that the promoter caspase will be activated by cleavage.

As mentioned above, there are numerous members of the caspases family, most of which are involved in apoptosis, but not all of them mediate apoptosis [24]. For example, the first discovered caspase, human interleukin-l-converting enzyme (ICE), was not associated with apoptosis but was responsible for mediating the inflammatory response. After the discovery of ICE, similar proteins to ICE were identified in *C. elegans* and were confirmed to be involved in the apoptotic process.

2.3 Apoptosis and peroxisomes

Peroxisomes, similar to the mitochondria, are a membranous subcellular organelle within eukaryotic cells. The peroxisome contains enzymes related to fatty acid and amino acid oxidation processes that produce hydrogen peroxide and also degrade hydrogen peroxide [25]. This gives the peroxisome its name and it plays an important role in maintaining intracellular oxidative metabolic homeostasis.

Because of the crucial role of the peroxisome, its dysfunction is associated with various pathological conditions, organ dysfunction, and aging [26–28]. For example, deficiency of Pex3, a peroxisomal membrane protein essential for membrane assembly, a member of the peroxisome (Pex) family, leads to complete loss of peroxisome function, while deficiency of Pex5, a peroxisome transporter, leads to Pex5 (a peroxisomal transporter) leads to the loss of peroxisomal matrix proteins. Mutations in this class of Pex genes may lead to human developmental abnormalities, such as human autosomal recessive disorders [29].

Peroxisomes play important roles in biosynthesis and signal transduction, which cannot be achieved without interaction with other organelles in the cell. In particular, peroxisomes interact functionally with mitochondria [30]. They cooperate with each other to perform biological functions such as production, fission, proliferation and degradation through vesicular transport, signaling, and membrane contact [31]. On the other hand, they can act synergistically to clear excess intracellular ROS, resist extracellular stresses through immune responses, and play an important role in the maintenance of lipid homeostasis through fatty acid β-oxidation [32–34]. In one word, peroxisomes are essential for the maintenance of normal mitochondrial and even whole cell function. Some chemotherapeutic drugs have been found to trigger mitochondrial dysfunction, leading to apoptosis by overwhelming cells with ROS. For example, Vorinostat (Vor), an FDA-approved histone deacetylase inhibitor (HDACi) for lymphoma treatment, has been well documented to trigger mitochondrial-mediated apoptosis through ROS accumulation. Acute Vor treatment has been shown to induce the expression of peroxisome proteins, thereby increasing peroxisome proliferation in a lymphoma model system. In addition, the knockdown of peroxisomes by gene silencing of Pex3 enhances Vor-induced ROS-mediated apoptosis [35].

In short, peroxisome dysfunction severely affects mitochondrial metabolism, cellular morphological stability, and biosynthesis, directly or indirectly contributing to a number of apoptosis-related diseases such as cancer [36, 37], cardiovascular disease [38–40], and neurodegenerative disorders [41].

2.4 Apoptosis-related diseases and peroxisomes

Apoptosis is an important way for the organism to maintain the numerical homeostasis of the cell population. Excessive or insufficient apoptosis can lead to disease.

2.4.1 Cancer

Crosstalk between mitochondria and other organelles is important in tumorigenesis. Mitochondria and peroxisomes are important organelles for ROS production and scavenging. Under normal conditions, both maintain intracellular ROS homeostasis. Impaired peroxisome function inevitably leads to increased levels of ROS in mitochondria, which impairs mitochondria, exacerbates impaired ROS clearance, leads to low levels of apoptosis, and thus promotes tumorigenesis and progression [42–44].

ROS act as signaling molecules to regulate various physiological and pathological processes [45]. H2O2 is a member of the ROS family and plays an important role in the signaling of epidermal growth factor (EGF) and platelet-derived growth factor (PDGF). H2O2 prevents protein tyrosine phosphatase 1B (PTP1B) from dephosphorylating EGF, thereby facilitating EGF stimulation. In addition, activation of PDGF requires H2O2 to promote oxidation and inactivation of PDGF-receptor-associated phosphatases and SHP-2, thereby facilitating the signaling pathway [46, 47]. Excessive ROS production can lead to cellular genomic instability (including mutations in the mitochondrial genome) on the one hand. Notably, ROS can promote tumor cell proliferation under hypoxic conditions. The reason for this is that the transcription factors hypoxia-inducible factors (HIFs) are upregulated under hypoxic conditions, thus promoting the expression of oncogenes. Although some proteases such as prolyl hydroxylases (PHDs) can degrade HIFs, the increased release of ROS induced by hypoxia can prevent the action of PHDs on HIFs. In this case, HIFs can then promote tumor progression under hypoxic conditions.

Briefly, because disruption of the functional balance between mitochondria and peroxidases may lead to increased ROS production, the increased ROS may inhibit apoptosis-inducing genes (bcl2 and p53, etc.), resulting in non-apoptosis of cells that should be apoptotic. Alternatively, the apoptotic process may be inhibited due to a decrease in the activity of apoptosis-related enzymes (caspases, etc.), leading to malignant cell transformation and tissue malignant proliferation. Both of these aspects are considered to be one of the important mechanisms leading to tumorigenesis and infiltrative metastasis.

2.4.2 Cardiovascular diseases (CVDs)

Apoptosis is a form of death of terminally differentiated cardiomyocytes. Clinical data suggest that ROS generation, DNA damage, and other factors activate apoptosis, resulting in the loss of large numbers of cardiomyocytes in patients with advanced congestive heart failure, patients with myocardial infarction, and patients with diabetic cardiomyopathy. The evidence suggests that apoptosis may be an important pathogenetic mechanism in cardiovascular disease [38]. Apoptosis, in concert with necrosis, may also lead to foam cell death and thus to the formation of a necrotic core, which contributes to lesion instability and increases the risk of lesion rupture and thrombosis.

Lower levels of ROS production can lead to chronic remodeling of the heart, whereas high levels of ROS can directly lead to apoptosis in the cardiomyocytes [48]. It is therefore interesting that catalase overexpression inhibits cardiomyocyte apoptosis by protecting the cells from ROS [49]. Peroxisomal antioxidant enzymes and plasmalogens protect cardiomyocytes via the degradation and trapping of ROS and the maintenance of ROS homeostasis. Apoptosis of cardiac cells has been demonstrated in several cardiovascular diseases, including myocardial ischemia–reperfusion injury (I/R) and atherosclerosis [50–52]. Atherosclerosis, a major cause of heart failure and myocardial infarction, can likewise predispose to acute coronary heart disease. There is evidence that thrombosis and plaque rupture may be due to apoptosis of a large number of smooth muscle cells and macrophages in unstable atherosclerotic plaques [53, 54]. Rupture of atherosclerotic plaques with concomitant thrombus formation may lead to coronary artery occlusion, which affects the blood supply to the myocardium, resulting in myocardial infarction and leading to patient death. Reperfusion is an effective treatment for acute myocardial infarction, but it may cause reperfusion injury while restoring blood flow [55]. Studies in the last decade or so have shown that cardiac cell death occurring during reperfusion after myocardial infarction is mainly apoptosis, not cell necrosis, which breaks the long-held misconception [56–58]. Usually, what occurs during I/R is mostly cell apoptosis, whereas necrosis occurs more often after prolonged ischemia. In addition, apoptosis also plays an important role in myocardial remodeling after infarction. There is evidence that a large number of apoptotic cells can be detected in myocardium at the marginal zone of myocardial infarction [56]. Since the regenerative capacity of myocardium is limited, people show great interest in preventing apoptosis of myocardial cells during I/R.

There is also a connection between chronic heart failure and apoptosis [59]. It has been reported that patients with advanced heart failure have higher rates of cardiac myocyte apoptosis than normal subjects. Using transgenic mice with cardiac tissue-specific expression of caspase-8, it was found that apoptosis of cardiomyocytes, even at very low levels, can lead to fatal dilated cardiomyopathy as long as it occurs chronically [60]. In addition, the use of caspase inhibitors prevented left ventricular dilatation and improved ventricular function, suggesting that long-term apoptosis can lead to a significant reduction in cardiomyocyte numbers, which in turn gradually decreases cardiac contractile function. As a result, the remaining cardiomyocytes become overcompensated and contribute to cardiac hypertrophy, leading to the development of heart failure [61].

Regarding the major pathways involved in apoptotic signaling in the heart, the death receptor pathway, the mitochondrial, and ER-stress death pathways are all involved [62]. The cross-talk between death receptors and mitochondrial cell death pathways has been demonstrated in cardiomyocytes and the heart [63, 64].

For example, Date *et al.* found that overexpression of FasL of the death receptor pathway activated both caspase-8 and -9 in cardiac myocytes [65]. Cardiac-restricted overexpression of TNF-α promoted apoptosis, but when these mice overexpressing TNF-α were crossed with mice overexpressing Bcl-2 in the heart, both left ventricular remodeling and cardiac apoptosis in the progeny mice were be alleviated [66].

In recent years, ER stress pathway has been reported to be in cross-talk with both the death receptor pathway and the mitochondrial pathway [13, 67, 68]. One study found that application of TNF-α induced HL-1 myoblast cell lines that activated both caspase-3 and -12 [69]. Bcl-2, which targets ER, inhibited mitochondrial membrane depolarization in apoptotic cells and also inhibited cytochrome c release [70]. Caspase-8 cleaves BAP31, an ER-associated protein, and the cleaved fragment induces Ca^{2+} release from ER, into the mitochondria, and initiates apoptosis [71]. It has also been reported that Bik proteins can activate Bax/Bak in the ER membrane after localization to the mitochondria, initiating Ca^{2+} release [72].

Regardless of the causative factor, and regardless of which signal transduction pathway or pathways are involved, oxidative stress due to the interaction of peroxisomes and mitochondria plays a pivotal role in triggering apoptosis and thus contributing to the development of cardiovascular disease.

2.4.3 Neurological disorders

Apoptosis plays a key role in the normal development of the central nervous system and is involved in the pathogenesis of adult brain-related diseases, such as stroke [73] and neurodegenerative diseases [74].

There is growing evidence that the decline in peroxisome function with age may be associated with age-related neurodegenerative diseases such as Alzheimer's disease (AD) and Parkinson's disease (PD) [75]. In the brains of patients with Alzheimer's disease and Parkinson's disease, plasmin levels are significantly reduced [76, 77], which suggests peroxisome dysfunction in neurodegenerative diseases. The lack of peroxisome activity in aged cells accumulates cellular ROS, which can compromise the integrity of organelles including mitochondria and the peroxisome itself. Subsequent defects in energy production mediated by peroxisomal fatty acid metabolism and mitochondrial oxidative phosphorylation may lead to metabolic failure in aged postmitotic cells, thereby inducing apoptosis associated with neurodegeneration.

Huntington's disease (HD), a prototypical neurodegenerative disorder, is caused by a mutation in the Huntingtin protein due to a repeat amplification of the CAG in the Huntington gene. Patients with this disease suffer from neuronal dysfunction due to massive apoptosis of nerve cells, which in turn manifests as mental cognitive and motor impairment, and even disability [74].

ROS can easily poison neurons due to their series of characteristics, such as rich in fatty acids, easy intracellular production of large amounts of hydroxyl radicals, weak antioxidant capacity, and low regenerative capacity. In addition, because of the high metabolic rates, neurons require a high energy supply from mitochondria, which are both the most important intracellular organelle for ROS production and also vulnerable to ROS attack. It has been shown that treatment of isolated cultured cerebellar granule neurons with hydrogen peroxide induces mitochondrial fission within 1 hour [78]. Furthermore, treatment of mice with nitric oxide in stroke leads to massive fission of neuronal mitochondria before the onset of neuronal loss [79]. In the presence of calcium, acute exposure to high levels of ROS can induce massive opening of mitochondrial membrane transition pores and increased permeability, which in turn causes cell Apoptosis or necrosis occurs. ROS production in mitochondria forms a vicious cycle with oxidative stress and is toxic to cells. There is some evidence in transgenic mouse models of HD that showed that Tauroursodeoxycholic acid (TUDCA), a hydrophilic bile acid with antioxidant properties, prevents the production of reactive oxygen species, mitigates mitochondrial insufficiency and apoptosis, in part, by inhibiting Bax translocation from cytosol to the mitochondria [80]. TUDCA prevented striatal degeneration and ameliorated locomotor and cognitive deficits in a 3-NP (3-nitropropionic acid) rat model of HD. Keene *et al.* [81] showed that systemically administered TUDCA significantly reduced striatal neuropathology, decreased striatal apoptosis, reduced the size of ubiquitinated neuronal intranuclear htt inclusions, and improved locomotor and sensorimotor deficits in the R6/2 transgenic HD mouse.

3. Conclusions

Apoptosis is a highly regulated cell death program that can be induced by a variety of physiological and pathological factors and has specific morphological and biochemical characteristics. The mechanism of its onset has not been completely elucidated to date, and it is now accepted that it is mediated by a number of pathways including the death receptor signaling pathway, the mitochondrial signaling pathway, and the endoplasmic reticulum signaling pathway. As an important way for the organism to maintain the numerical homeostasis of the cell population, apoptosis plays a key role in the pathogenesis of various human diseases. Peroxisomes and mitochondria are membrane-bound organelles in the cytoplasm of eukaryotic cells and are closely related to each other in their organelle synthesis and function. One of their important roles in cooperating with each other is to regulate the level and extent of apoptosis by maintaining the homeostasis of reactive oxygen species in the cell. Peroxisome dysfunction severely affects mitochondrial metabolism, cellular morphological stability, and biosynthesis, and therefore contributes directly or indirectly to a number of apoptosis-related diseases. Based on the available relevant findings, this chapter presents and summarizes the important potential role of peroxisomes in apoptosis-related diseases such as tumors, cardiovascular diseases, and neuropsychiatric disorders.

Acknowledgements

This work was supported in part by grants from National Natural Science Foundation of China (22176002), Anhui Provincial Natural Science Foundation (2008085 MB49), Natural Science Foundation of Anhui Provincial Department of Education (KJ2021A0215), Anhui Medical University Research Enhancement Program (2021xkjT004), and Open Project Fund of the Key Laboratory of the Ministry of Education for the Birth Population (JKZD20202).

Conflict of interest

The authors report no conflicts of interest.

Author details

Meimei Wang[1*], Yakun Liu[1], Ni Chen[2], Juan Wang[3] and Ye Zhao[2]

1 Department of Pathophysiology, School of Basic Medical Science, Anhui Medical University, Hefei, Anhui, P.R. China

2 Department of Nuclear Medicine, School of Basic Medical Science, Anhui Medical University, Hefei, Anhui, P.R. China

3 Department of Public Health Inspection and Quarantine, School of Public Health, Anhui Medical University, Hefei, Anhui, P.R. China

*Address all correspondence to: wangmm@ustc.edu.cn

References

[1] Galluzzi L et al. Molecular mechanisms of cell death: Recommendations of the nomenclature committee on cell death 2018. Cell Death and Differentiation. 2018;**25**:486-541. DOI: 10.1038/s41418-017-0012-4

[2] Tummers B, Green DR. The evolution of regulated cell death pathways in animals and their evasion by pathogens. Physiological Reviews. 2022;**102**:411-454. DOI: 10.1152/physrev.00002.2021

[3] Wang Y, Kanneganti TD. From pyroptosis, apoptosis and necroptosis to PANoptosis: A mechanistic compendium of programmed cell death pathways. Computational and Structural Biotechnology Journal. 2021;**19**:4641-4657. DOI: 10.1016/j.csbj.2021.07.038

[4] Fuchs Y, Steller H. Live to die another way: Modes of programmed cell death and the signals emanating from dying cells. Nature Reviews. Molecular Cell Biology. 2015;**16**:329-344. DOI: 10.1038/nrm3999

[5] Pérez-Garijo A, Steller H. Spreading the word: Non-autonomous effects of apoptosis during development, regeneration and disease. Development. 2015;**142**:3253-3262. DOI: 10.1242/dev.127878

[6] Ishizuya-Oka A, Hasebe T, Shi YB. Apoptosis in amphibian organs during metamorphosis. Apoptosis. 2010;**15**:350-364. DOI: 10.1007/s10495-009-0422-y

[7] Nirmala JG, Lopus M. Cell death mechanisms in eukaryotes. Cell Biology and Toxicology. 2020;**36**:145-164. DOI: 10.1007/s10565-019-09496-2

[8] Proskuryakov SY, Konoplyannikov AG, Gabai VL. Necrosis: A specific form of programmed cell death? Experimental Cell Research. 2003;**283**:1-16. DOI: 10.1016/s0014-4827(02)00027-7

[9] Aruscavage PJ, Hellwig S, Bass BL. Small DNA pieces in C. elegans are intermediates of DNA fragmentation during apoptosis. PLoS One. 2010;**5**:e11217. DOI: 10.1371/journal.pone.0011217

[10] Tyurin VA et al. Oxidatively modified phosphatidylserines on the surface of apoptotic cells are essential phagocytic 'eat-me' signals: Cleavage and inhibition of phagocytosis by Lp-PLA2. Cell Death and Differentiation. 2014;**21**:825-835. DOI: 10.1038/cdd.2014.1

[11] Liu Y et al. Cell-type apoptosis in lung during SARS-CoV-2 infection. Pathogens. 2021;**10**:509. DOI: 10.3390/pathogens10050509

[12] Mandal P, McCormick AL, Mocarski ES. TNF Signaling dictates myeloid and non-myeloid cell crosstalk to execute MCMV-induced extrinsic apoptosis. Viruses. 2020;**12**:1221. DOI: 10.3390/v12111221

[13] Kim NH, Kang PM. Apoptosis in cardiovascular diseases: Mechanism and clinical implications. Korean Circulation Journal. 2010;**40**:299-305. DOI: 10.4070/kcj.2010.40.7.299

[14] Sp N et al. Tannic acid promotes TRAIL-induced extrinsic apoptosis by regulating mitochondrial ROS in human embryonic carcinoma cells. Cells. 2020;**9**:282. DOI: 10.3390/cells9020282

[15] Ranjan K, Waghela BN, Vaidya FU, Pathak C. Cell-penetrable peptide-conjugated FADD induces apoptosis

and regulates inflammatory Signaling in Cancer cells. International Journal of Molecular Sciences. 2020;**21**:6890. DOI: 10.3390/ijms21186890

[16] Dowling JP, Nair A, Zhang J. A novel function of RIP1 in postnatal development and immune homeostasis by protecting against RIP3-dependent necroptosis and FADD-mediated apoptosis. Frontiers in Cell and Development Biology. 2015;**3**:12. DOI: 10.3389/fcell.2015.00012

[17] Miao X et al. IAP-1 promoted cisplatin resistance in nasopharyngeal carcinoma via inhibition of caspase-3-mediated apoptosis. American Journal of Cancer Research. 2021;**11**:640-667

[18] Green DR, Reed JC. Mitochondria and apoptosis. Science. 1998;**281**:1309-1312. DOI: 10.1126/science.281.5381.1309

[19] Kopeina GS, Prokhorova EA, Lavrik IN, Zhivotovsky B. Alterations in the nucleocytoplasmic transport in apoptosis: Caspases lead the way. Cell Proliferation. 2018;**51**:e12467. DOI: 10.1111/cpr.12467

[20] Ueda N. Ceramide-induced apoptosis in renal tubular cells: A role of mitochondria and sphingosine-1-phoshate. International Journal of Molecular Sciences. 2015;**16**:5076-5124. DOI: 10.3390/ijms16035076

[21] Singh R, Letai A, Sarosiek K. Regulation of apoptosis in health and disease: The balancing act of BCL-2 family proteins. Nature Reviews. Molecular Cell Biology. 2019;**20**:175-193. DOI: 10.1038/s41580-018-0089-8

[22] Zheng P et al. DNA damage triggers tubular endoplasmic reticulum extension to promote apoptosis by facilitating ER-mitochondria signaling. Cell Research. 2018;**28**:833-854. DOI: 10.1038/s41422-018-0065-z

[23] Kim C, Kim B. Anti-Cancer natural products and their bioactive compounds inducing ER stress-mediated apoptosis: A review. Nutrients. 2018;**10**:1021. DOI: 10.3390/nu10081021

[24] Sinderewicz E et al. Expression of factors involved in apoptosis and cell survival is correlated with enzymes synthesizing lysophosphatidic acid and its receptors in granulosa cells originating from different types of bovine ovarian follicles. Reproductive Biology and Endocrinology. 2017;**15**:72. DOI: 10.1186/s12958-017-0287-9

[25] Islinger M, Voelkl A, Fahimi HD, Schrader M. The peroxisome: An update on mysteries 2.0. Histochemistry and Cell Biology. 2018;**150**:443-471. DOI: 10.1007/s00418-018-1722-5

[26] Manivannan S, Scheckhuber CQ, Veenhuis M, van der Klei IJ. The impact of peroxisomes on cellular aging and death. Frontiers in Oncology. 2012;**2**:50. DOI: 10.3389/fonc.2012.00050

[27] Kim J, Bai H. Peroxisomal stress response and inter-organelle communication in cellular homeostasis and aging. Antioxidants. 2022;**11**:192. DOI: 10.3390/antiox11020192

[28] Marzetti E et al. Mitochondrial dysfunction and sarcopenia of aging: From signaling pathways to clinical trials. The International Journal of Biochemistry & Cell Biology. 2013;**45**:2288-2301. DOI: 10.1016/j.biocel.2013.06.024

[29] Uzor NE, McCullough LD, Tsvetkov AS. Peroxisomal dysfunction in neurological diseases and brain aging. Frontiers in Cellular Neuroscience. 2020;**14**:44. DOI: 10.3389/fncel.2020.00044

[30] Lismont C, Nordgren M, Van Veldhoven PP, Fransen M. Redox interplay between mitochondria and peroxisomes. Frontiers in Cell and Development Biology. 2015;**3**:35. DOI: 10.3389/fcell.2015.00035

[31] Tang Y et al. LncRNAs regulate the cytoskeleton and related rho/ROCK signaling in cancer metastasis. Molecular Cancer. 2018;**17**:77. DOI: 10.1186/s12943-018-0825-x

[32] Cohen Y et al. Peroxisomes are juxtaposed to strategic sites on mitochondria. Molecular BioSystems. 2014;**10**:1742-1748. DOI: 10.1039/c4mb00001c

[33] Hosoi KI et al. The VDAC2-BAK axis regulates peroxisomal membrane permeability. The Journal of Cell Biology. 2017;**216**:709-722. DOI: 10.1083/jcb.201605002

[34] Schrader M, Costello J, Godinho LF, Islinger M. Peroxisome-mitochondria interplay and disease. Journal of Inherited Metabolic Disease. 2015;**38**:681-702. DOI: 10.1007/s10545-015-9819-7

[35] Dahabieh MS et al. Peroxisomes protect lymphoma cells from HDAC inhibitor-mediated apoptosis. Cell Death and Differentiation. 2017;**24**:1912-1924. DOI: 10.1038/cdd.2017.115

[36] Dahabieh MS et al. Peroxisomes and cancer: The role of a metabolic specialist in a disease of aberrant metabolism. Biochimica Et Biophysica Acta. Reviews on Cancer. 2018;**1870**:103-121. DOI: 10.1016/j.bbcan.2018.07.004

[37] Kaul D. Cholesterol-receptor-mediated genomics in health and disease. Trends in Molecular Medicine. 2003;**9**:442-449. DOI: 10.1016/j.molmed.2003.08.010

[38] Lee Y, Gustafsson AB. Role of apoptosis in cardiovascular disease. Apoptosis. 2009;**14**:536-548. DOI: 10.1007/s10495-008-0302-x

[39] van Bilsen M, van Nieuwenhoven FA. PPARs as therapeutic targets in cardiovascular disease. Expert Opinion on Therapeutic Targets. 2010;**14**:1029-1045. DOI: 10.1517/14728222.2010.512917

[40] Gerry JM, Pascual G. Narrowing in on cardiovascular disease: The atheroprotective role of peroxisome proliferator-activated receptor gamma. Trends in Cardiovascular Medicine. 2008;**18**:39-44. DOI: 10.1016/j.tcm.2007.12.001

[41] Grabowski GA. Overview of inflammation in Neurometabolic diseases. Seminars in Pediatric Neurology. 2017;**24**:207-213. DOI: 10.1016/j.spen.2017.08.005

[42] Kim J-A. Peroxisome metabolism in Cancer. Cells. 2020;**9**:1692. DOI: 10.3390/cells9071692

[43] Tanaka H et al. Peroxisomes control mitochondrial dynamics and the mitochondrion-dependent apoptosis pathway. Journal of Cell Science. 2019;**132**:jcs224766. DOI: 10.1242/jcs.224766

[44] Xia M et al. Communication between mitochondria and other organelles: A brand-new perspective on mitochondria in cancer. Cell & Bioscience. 2019;**9**:27. DOI: 10.1186/s13578-019-0289-8

[45] Reczek CR, Chandel NS. ROS-dependent signal transduction. Current Opinion in Cell Biology. 2015;**33**:8-13. DOI: 10.1016/j.ceb.2014.09.010

[46] Schieber M, Chandel NS. ROS function in redox signaling and

oxidative stress. Current Biology. 2014;**24**:R453-R462. DOI: 10.1016/j.cub.2014.03.034

[47] Diebold L, Chandel NS. Mitochondrial ROS regulation of proliferating cells. Free Radical Biology & Medicine. 2016;**100**:86-93. DOI: 10.1016/j.freeradbiomed.2016.04.198

[48] Sugamura K, Keaney JF. Reactive oxygen species in cardiovascular disease. Free Radical Biology & Medicine. 2011;**51**:978-992. DOI: 10.1016/j.freeradbiomed.2011.05.004

[49] Qin F et al. Cardiac-specific overexpression of catalase identifies hydrogen peroxide-dependent and -independent phases of myocardial remodeling and prevents the progression to overt heart failure in G(alpha)q-overexpressing transgenic mice. Circulation. Heart Failure. 2010;**3**:306-313. DOI: 10.1161/CIRCHEARTFAILURE.109.864785

[50] Neve BP, Fruchart JC, Staels B. Role of the peroxisome proliferator-activated receptors (PPAR) in atherosclerosis. Biochemical Pharmacology. 2000;**60**:1245-1250. DOI: 10.1016/s0006-2952(00)00430-5

[51] Puddu P, Puddu GM, Muscari A. Peroxisome proliferator-activated receptors: Are they involved in atherosclerosis progression? International Journal of Cardiology. 2003;**90**:133-140. DOI: 10.1016/s0167-5273(02)00565-x

[52] Gao Z et al. Mechanistic insight into PPARγ and Tregs in atherosclerotic immune inflammation. Frontiers in Pharmacology. 2021;**12**:750078. DOI: 10.3389/fphar.2021.750078

[53] Kockx MM et al. Apoptosis and related proteins in different stages of human atherosclerotic plaques. Circulation. 1998;**97**:2307-2315. DOI: 10.1161/01.cir.97.23.2307

[54] Clarke MC et al. Chronic apoptosis of vascular smooth muscle cells accelerates atherosclerosis and promotes calcification and medial degeneration. Circulation Research. 2008;**102**:1529-1538. DOI: 10.1161/CIRCRESAHA.108.175976

[55] Matsumura K, Jeremy RW, Schaper J, Becker LC. Progression of myocardial necrosis during reperfusion of ischemic myocardium. Circulation. 1998;**97**:795-804. DOI: 10.1161/01.cir.97.8.795

[56] Olivetti G et al. Acute myocardial infarction in humans is associated with activation of programmed myocyte cell death in the surviving portion of the heart. Journal of Molecular and Cellular Cardiology. 1996;**28**:2005-2016. DOI: 10.1006/jmcc.1996.0193

[57] Saraste A et al. Apoptosis in human acute myocardial infarction. Circulation. 1997;**95**:320-323. DOI: 10.1161/01.cir.95.2.320

[58] Gottlieb RA, Burleson KO, Kloner RA, Babior BM, Engler RL. Reperfusion injury induces apoptosis in rabbit cardiomyocytes. The Journal of Clinical Investigation. 1994;**94**:1621-1628. DOI: 10.1172/JCI117504

[59] Gurusamy N, Das DK. Is autophagy a double-edged sword for the heart? Acta Physiologica Hungarica. 2009;**96**:267-276. DOI: 10.1556/APhysiol.96.2009.3.2

[60] Abushouk AI et al. Peroxisome proliferator-activated receptors as therapeutic targets for heart failure. Biomedicine & Pharmacotherapy. 2017;**95**:692-700. DOI: 10.1016/j.biopha.2017.08.083

[61] Wencker D et al. A mechanistic role for cardiac myocyte apoptosis in heart failure. The Journal of Clinical Investigation. 2003;**111**:1497-1504. DOI: 10.1172/JCI17664

[62] Gutiérrez-Cuevas J, Santos A, Armendariz-Borunda J. Pathophysiological molecular mechanisms of obesity: A link between MAFLD and NASH with cardiovascular diseases. International Journal of Molecular Sciences. 2021;**22**:11629. DOI: 10.3390/ijms222111629

[63] Luo X, Budihardjo I, Zou H, Slaughter C, Wang X. Bid, a Bcl2 interacting protein, mediates cytochrome c release from mitochondria in response to activation of cell surface death receptors. Cell. 1998;**94**:481-490. DOI: 10.1016/s0092-8674(00)81589-5

[64] Li H, Zhu H, Xu CJ, Yuan J. Cleavage of BID by caspase 8 mediates the mitochondrial damage in the Fas pathway of apoptosis. Cell. 1998;**94**:491-501. DOI: 10.1016/s0092-8674(00)81590-1

[65] Date T et al. Differential effects of membrane and soluble Fas ligand on cardiomyocytes: Role in ischemia/reperfusion injury. Journal of Molecular and Cellular Cardiology. 2003;**35**:811-821. DOI: 10.1016/s0022-2828(03)00139-1

[66] Haudek SB, Taffet GE, Schneider MD, Mann DL. TNF provokes cardiomyocyte apoptosis and cardiac remodeling through activation of multiple cell death pathways. The Journal of Clinical Investigation. 2007;**117**:2692-2701. DOI: 10.1172/JCI29134

[67] Sugiura A, Mattie S, Prudent J, McBride HM. Newly born peroxisomes are a hybrid of mitochondrial and ER-derived pre-peroxisomes. Nature. 2017;**542**:251-254. DOI: 10.1038/nature21375

[68] Kalai M et al. Regulation of the expression and processing of caspase-12. The Journal of Cell Biology. 2003;**162**:457-467. DOI: 10.1083/jcb.200303157

[69] Bajaj G, Sharma RK. TNF-alpha-mediated cardiomyocyte apoptosis involves caspase-12 and calpain. Biochemical and Biophysical Research Communications. 2006;**345**:1558-1564. DOI: 10.1016/j.bbrc.2006.05.059

[70] Gyan E et al. Spontaneous and Fas-induced apoptosis of low-grade MDS erythroid precursors involves the endoplasmic reticulum. Leukemia. 2008;**22**:1864-1873. DOI: 10.1038/leu.2008.172

[71] Breckenridge DG, Stojanovic M, Marcellus RC, Shore GC. Caspase cleavage product of BAP31 induces mitochondrial fission through endoplasmic reticulum calcium signals, enhancing cytochrome c release to the cytosol. The Journal of Cell Biology. 2003;**160**:1115-1127. DOI: 10.1083/jcb.200212059

[72] Mathai JP, Germain M, Shore GC. BH3-only BIK regulates BAX,BAK-dependent release of Ca2+ from endoplasmic reticulum stores and mitochondrial apoptosis during stress-induced cell death. The Journal of Biological Chemistry. 2005;**280**:23829-23836. DOI: 10.1074/jbc.M500800200

[73] Tatton WG, Olanow CW. Apoptosis in neurodegenerative diseases: The role of mitochondria. Biochimica et Biophysica Acta. 1999;**1410**:195-213. DOI: 10.1016/s0005-2728(98)00167-4

[74] Zhang H et al. Glucagon-like peptide-1 attenuated carboxymethyl lysine induced neuronal apoptosis via

peroxisome proliferation activated receptor-γ. Aging. 2021;**13**:19013-19027. DOI: 10.18632/aging.203351

[75] Cipolla CM, Lodhi IJ. Peroxisomal dysfunction in age-related diseases. Trends in Endocrinology and Metabolism. 2017;**28**:297-308. DOI: 10.1016/j.tem.2016.12.003

[76] Kou J et al. Peroxisomal alterations in Alzheimer's disease. Acta Neuropathologica. 2011;**122**:271-283. DOI: 10.1007/s00401-011-0836-9

[77] Fabelo N et al. Severe alterations in lipid composition of frontal cortex lipid rafts from Parkinson's disease and incidental Parkinson's disease. Molecular Medicine. 2011;**17**:1107-1118. DOI: 10.2119/molmed.2011.00119

[78] Jahani-Asl A et al. Mitofusin 2 protects cerebellar granule neurons against injury-induced cell death. The Journal of Biological Chemistry. 2007;**282**:23788-23798. DOI: 10.1074/jbc.M703812200

[79] Barsoum MJ et al. Nitric oxide-induced mitochondrial fission is regulated by dynamin-related GTPases in neurons. The EMBO Journal. 2006;**25**:3900-3911. DOI: 10.1038/sj.emboj.7601253

[80] Johri A, Beal MF. Antioxidants in Huntington's disease. Biochimica et Biophysica Acta. **1822;664-674**:2012. DOI: 10.1016/j.bbadis.2011.11.014

[81] Keene CD et al. Tauroursodeoxycholic acid, a bile acid, is neuroprotective in a transgenic animal model of Huntington's disease. Proceedings of the National Academy of Sciences of the United States of America. 2002;**99**:10671-10676. DOI: 10.1073/pnas.162362299

Chapter 4

Association of Peroxisomes, Reactive Oxygen Species (ROS) and Antioxidants: Insights from Preclinical and Clinical Evaluations

Nishat Fatima

Abstract

Peroxisome function has long been associated with oxygen metabolism. High concentrations of hydrogen peroxide (H_2O_2−) producing oxidases are in the set of peroxisomes and their antioxidant enzymes, especially catalase. Reactive oxygen species (ROS) can certainly be considered as an intracellular multifunctional biological factor which are released and scevenged in peroxisomes. They are known to be involved in normal cellular functions such as signaling mediators, overproduction under oxidative stress conditions leading to adverse cellular effects, cell death, and various other pathological conditions. This review provides an insight into the relationship between peroxisomes and ROS, which are emerging as key players in the dynamic rotation of ROS metabolism and oxidative damage. Various conditions upset the balance between ROS production and removal in peroxisomes. The current review also targets the ROS-inhibiting enzymes and exemplifying the effects of antioxidants in pre-clinical and clinical evaluation of natural and herbal supplements.

Keywords: antioxidants, oxidative stress, free radicals, vitamin C, E

1. Introduction

Reactive oxygen species (ROS) can certainly be considered as an intracellular multifunctional biological factor. They are known to be involved in normal cellular functions such as signaling mediators, overproduction under oxidative stress conditions leading to adverse cellular effects and eventually cell death. Under normal conditions in every human cell, the release of pro-oxidants in the form of ROS (reactive oxigen species) and RNS (reactive nitrogen species) are scrutinized by antioxidant levels. The equilibrium maintained is shifted in favor of pro-oxidants resulting in oxidative stress, when exposed to adverse circumstances such as atmospheric pollutants, unfavorable physicochemical, environmental or pathologicl agents including cigarette smoking, toxic chemicals, ultra violet rays and radiation and also excess formation of advanced glycation end products (AGE), in diabetes [1, 2]. This has been associated in the origin of various (>100) human diseases. Peroxisome function has long been associated with oxygen metabolism. High concentrations of $H_2O_2^-$ producing oxidases are in the set of peroxisomes and their antioxidant enzymes, especially catalase. This review provides an insight into the relationship between ROS and peroxisomes, which have emerged as key players in the dynamic rotation of ROS metabolism and oxidative damage. The counterparts of ROS-producing enzymes such as antioxidants are also discussed; exemplifying their effects in pre-clinical and clinical evaluation of natural and herbal supplements [3].

Peroxisomes contain at least 50 different enzymes involved in a variety of biochemical pathways in different cell types. Peroxisomes were originally defined as organelles that perform oxidation reactions that result in the production of hydrogen peroxide. Because hydrogen peroxide is harmful to the cell, peroxisomes also contain the enzyme catalase, which breaks down hydrogen peroxide either by converting it to water or by oxidizing another organic compound. A variety of substrates are degraded by such oxidative reactions in peroxisomes, including uric acid, amino acids, and fatty acids. The oxidation of fatty acids is a particularly important example as it is a major source of metabolic energy. In animal cells, fatty acids are oxidized in both peroxisomes and mitochondria, but in yeast and plants, fatty acid oxidation is restricted to peroxisomes. Peroxisomes are one of the main sites in the cell where oxygen free radicals are both generated and scavenged. The balance between these two processes is believed to be of great importance for the proper functioning of cells and has been linked to aging and carcinogenesis. The peroxisome is a single, membrane-bound organelle present in virtually every eukaryotic cell and biosynthetic pathways, however, these pathways may differ between species. The significance of the peroxisome for the normal regulation of cellular activities can be explained by the fact that more than 20 human peroxisomal disorders dwell because of the absence of protein or due to loss of protein function [4].

Antioxidants act either by neutralizing free radicals or their consequences [5]. The natural cellular environment provides sufficient protective pathways against unfavorable effects of free radicals: glutathione reductase, glutathione peroxidase, superoxide dismutase (SOD), disulphide bonding, thiols and thioredoxin are buffering systems in every cell. The relation between free radicals and disease can be explained by the concept of "Oxidative stress" [3].

All biological molecules present in our body are at risk of being attacked by free radicals. Such damaged molecules can impair the cell functions and even lead to cell death eventually resulting in disease states. Antioxidants may prevent and improve different diseased states [1, 2]. Several investigators have demonstrated the positive effects of antioxidants like Vitamin C, E to name a few in both preclinical and clinical setup.

2. Oxidative stress as a marker of endothelial dysfunction

The vascular endothelium, which promotes the passage of macromolecules and circulating cells from blood to tissues, is an important target of oxidative stress, playing a pivotal role in the pathophysiology of several vascular diseases and disorders. It is has been also reported that exclusively, oxidative stress increases vascular endothelial permeability and promotes leukocyte adhesion, which is coupled with alterations in endothelial signal transduction and redox-regulated transcription factors [6]. The reactive oxygen species (ROS) which originate at the sites of inflammation and injury have been emerged as the major contributing factor in the pathogenesis of endothelial dysfunction. These reactive oxygen species at low concentrations can function as signaling molecules participating in the regulation of fundamental cell activities such as cell growth and cell adaptation responses; whereas at higher concentrations, ROS can cause cellular injury and death. Predominantly, under normal body functions, the vital enzymes such as NAD(P)H dependent oxidases and superoxide dismutases (SOD) conscientiously regulate release of superoxide along with maintenance of intracellular redox balance. On the other hand, if production of superoxide anion surpasses the scavenging capacity of endothelial cells, this active intermediate react with nitric oxide resulting in formation of peroxynitrites. These peroxynitrites are potent oxidants that cause structural and functional changes in various components of the cell. ROS are also well known to suppress NO and restrict the formation of peroxynitrite [7]. It is a cytotoxic oxidant that causes endotheial dysfunction through nitration of protein function. Peroxynitrite plays a key role in oxidation of LDL and as proatherogenic [8]. In addition, peroxynitrite leads to the degradation of the eNOS cofactor tetrahydrobiopterin (BH4), resulting in an uncoupling of eNOS [9]. An excess of oxidant also leads to a reduction of BH4 with an increase in BH2. When this occurs, the formation of the active dimer of eNOS with oxygenase activity and the production of NO is restricted. The reductase function of eNOS is activated and more ROS are produced, so NO synthase switches from its NO-producing oxygenase function to its ROS-producing reductase function, with consequent exaggeration of oxidant excess and deleterious effects on endothelial and vascular function of the vessel wall. ROS upregulate adhesion (VCAM-1 and ICAM-1) and chemotactic molecules (macrophage chemoattractant peptide-1 (MCP-1) [8]. Inflammation decreases the bioavailability of NO [8]. The main source of oxidative excess in the vasculature is NAD(P)H oxidase and xanthine oxidase [10], the mitochondria [11] and uncoupled NOS constitute as other sources.

3. Significance of antioxidants

Many investigators have studied the significance of antioxidants in relation to disease and showed that zinc is an essential trace element, being a cofactor for about 200 human enzymes, including cytosolic antioxidant Cu-Zn SOD, isoenzyme of SOD mainly present in cytosol. Selenium is also an essential trace element and a cofactor for glutathione peroxidase. There is a vast information which suggests that chronic administration of antioxidants may be beneficial in improving cardiovascular risk. Vitamin E and tocotrienols (such as those from palm oil) are efficient lipid soluble antioxidants that function as a chain breaker during lipid peroxidation in cell memebranes and various lipid particles including LDL [12].

Vitamin E is considered the standard antioxidant against which other compounds with antioxidant activity are compared, particularly in terms of its biological activity and clinical relevance. Daily dietary intake varies between 400 IU and 800 IU. Vitamin C is also another important water-soluble free radical scavenger. The recommended daily dose is 60 mg. Apart from these, carotenoids like beta-carotene, lycopene, lutein and other carotenoids act as important antioxidants and quench superoxide (O2.) and (ROO..) [12, 13].

The effects of short-term dietary supplementation of tomato juice (source of lycopene), vitamin E and vitamin C on susceptibility of LDL to oxidation and circulating levels of C-reactive protein (CRP) and cell adhesion molecules measured in patients with type 2 diabetes. In this study 57 patients with well controlled type 2 diabetes melliitus were randomized to receive tomato juice (500 mg/day), Vitamin E (800 U/day), Vitamin C (500 mg/day) or placebo treatment for 4 weeks. It was observed that lycopene and vitamin E were both associated with resistance of LDL to oxidation, but only Vitamin E showed a decrease in C-reactive protein. It was also found that levels of cell adhesion molecules and plasma glucose did not change significantly during the study. Thses investigators then suggested that these findings may be relevant to strategies aimed at reducing risk of myocardial infarction in patients with diabetes [14].

In another study, it was suggested that vitamin E (1,600 IU/day, 10 weeks) decreased the susceptibility of LDL to oxidation in comparison with placebo. Vitamin E had this effect in both bouyant and dense LDL subfractions. This protection occurred in an environment where glycemic indices did not change and protein glycation was unaffected. The hypothesis that endothelial function and LDL oxidation might be linked was advanced further by Pinkney et al. [15]. These investigators studied 46 patients with type 1 diabetes without nephropathy and compared the results to 39 controls using a 3-month, randomized, double-blind, placebo-controlled study of vitamin E, 500 IU/day. The results indicated that in the absence of changes in LDL oxidation, vitamin E intake enhanced flow mediated dilatation FMD in type I diabetics [16].

Another [17] study reported that intracellular Vitamin C levels are reduced in patients with type 1 diabetes, particularly those who are poorly controlled. Histologically, the microvascular lesions of scurvy bear a surprising resemblance to those seen in long-standing diabetes, making them an attractive therapeutic alternative. Many short-term studies have shown the beneficial effects of ascorbic acid on vascular function, particularly in smokers and after ingestion of high-fat meals. However, the effect of vitamin C is not chronically sustainable, at least in smokers. Based on these observations, hypothesized that the antioxidant vitamin C might enhance endothelium-dependent vasodilation in forearm resistance when tested [18]. These investigators studied 10 subjects with diabetes and 10 age-matched, nondiabetic control subjects. FBF was determined by venous occlusion plethysmography, and endothelium dependent vasodilatation. The results from this study indeed support the hypothesis that acute administration of vitamin C improves endothelial function associated with the diabetic state., however, no information on chronic effects can be found from this study. But still more research has to be taken up in exploring the possible use of these vitamins to prevent atherosclerosis and/or microvascular disease in patients with diabetes. A study showed for the first time effects of consumption of flavonoid rich dark chocolate on endothelial function, aortic stiffness, wave reflections and oxidant status in healthy adults [19].

According to world health organization, traditional medicines are widely used globally. Approximately 80% of the population of developing countries rely on traditional medicines for their primary health care needs [20–22]. These medicinal plants contain several phytochemicals such as Vitamins (A, C,E and K), carotenoids, terpenoids, flavonoids, polyphenols, alkaloids, tannins, saponins, enzymes and minerals etc. These phytochemicals possess antioxidant activities, which can be used in the treatment of multiple ailments [23]. Many herbs along with potent antioxidant activity also possess anti-inflammatory and cardioprotective properties and are used by patients with increased risk of cardiovascular morbidity and mortality. Thus it is necessary to through light on the beneficial effects of the herbs such as *Terminalia arjuna, Emblica officinalis, Withania somnifera, Boerhaavia diffusa* and *Ocimum sanctum etc.*

4. Preclinical studies with *Terminalia arjuna*

4.1 Antioxidant and anticancer activities

The effect of Terminalia arjuna aqueous extract on the antioxidant defense system in lymphoma-bearing AKR mice was examined. The antioxidant effects of T. arjuna were monitored through the activities of catalase, superoxide dismutase and glutathione S-transferase. These enzyme activities are low in lymphoma-bearing mice, indicating an impaired antioxidant defense system. Oral administration of different doses of aqueous extracts of T. arjuna caused a significant increase in the activities of antioxidant enzymes. Here, T. arjuna was found to downregulate anaerobic metabolism by inhibiting lactate dehydrogenase activity in lymphoma-bearing mice, which was increased in untreated cancerous mice. The results demonstrated the antioxidant effects of Terminalia arjuna aqueous extract, which may play a role in anticarcinogenic activity by reducing oxidative stress [24].

4.1.1 Cardio protective activity

Sumitra et al., demonstrated that Arjunolic acid, a new triterpene and a potent principle from the bark of *Terminalia arjuna* has been shown to produce significant cardiac protection in isoproterenol induced myocardial necrosis in rats and prevents decrease in the levels of super oxide dismutase, catalase and reduced glutathione. This study explains that Arjunolic acid at a dosage of 15 mg/kg body weight (Pre and post treatment) produces cardioprotective effect [25].

4.1.2 Antiplatelet activity

Some researchers have shown that oleanane-type triterpene glycosides designated as Termiarjunoside I and Termiarjunoside II isolated from stem bark of *Terminalia arjuna*, potently suppressed the release of nitric oxide and superoxide from isolated macrophages and also inhibited aggregation of platelets [26].

5. Clinical studies with *Terminalia arjuna*

5.1 Antioxidant and Cardioprotective activity

The antioxidant constituents in Terminalia arjuna are reported to reverse endothelial dysfunction in chronic smokers. The study was conducted with 18 healthy male smokers and an equal number of non-smokers of the same age. The baseline brachial artery reactivity test was done using high-frequency ultrasound according to the standard protocol under identical conditions to determine endothelium-dependent flow-mediated and endothelium-independent nitroglycerin-mediated dilatation. The two groups were matched for age, body mass index, blood pressure, serum cholesterol, mean resting vessel diameter, and flow velocities after occlusion. Smokers then received Terminalia arjuna (500 mg every 8 hours) or a matched placebo randomly in a double-blind, cross-over design for two weeks each, followed by repeated brachial artery reactivity studies to determine various parameters, including flow-mediated dilatation after each period. However, flow-mediated dilatation showed a significant improvement from baseline after Terminalia arjuna therapy. The study concluded that smokers have impaired endothelium-dependent but normal endothelium-independent vasodilation as determined by brachial artery reactivity studies. In addition, two weeks of Terminalia arjuna therapy resulted in significant regression of this endothelial abnormality in smokers [27].

The effect of *Terminalia arjuna* (500 mg 8 hourly) was evaluated in fifty-eight males with chronic stable angina (NYHA class II-III) and with isosorbide mononitrate (40 mg/daily) on treadmill exercise induced ischemia, or a matching placebo for one week each. A wash-out period of at least three days was observed between the groups in a randomized, double-blind, crossover design. The treadmill exercise test parameters improved significantly during therapy with both treatments compared to those with placebo [28].

5.1.1 E. officinalis

The fruits of *E. officinalis* (Amla) family: Euphorbiaceae, commonly known as Indian gooseberry is widely used in many of the indigenous medical preparations against a variety of disease conditions [29].

E. officinalis is considered as a rich source of a vitamin C, which plays an important role in scavenging free radicals. For many years the therapeutic potential of fruits of *E. officinalis* was attributed to their high content of ascorbic acid [30]. It was further determined through comprehensive, chromatographic, spectroscopic and crucial chemical analyzes that the antioxidant property is due to the low molecular weight hydrolyzable tannins of fresh fruit skin. These tannins, namely Emblicanin A, Emblicanin B, Pedunculagin and Punigluconin, have been found to provide protection against oxygen radical-induced hemolysis of rat peripheral erythrocytes [31]. Purification and fractionation process was conducted in another study and phytochemicals like gallic acid, methyl gallate and geranin were identfied [32].

6. Preclinical studies with *E. officinalis*

6.1 Antioxidant activity

Invitro and animal studies have shown that Amla has potent antioxidant activity against multiple test systems such as superoxide radicals, induction of lipid peroxide formation by the Fe^{+3}/ADP ascorbate system, hydroxyl radical scavenging activity. It also caused systemic increase in antioxidant enzymes in laboratory animals [33].

6.2 Hypolipidemic activity

In a study conducted in rats showed that flavonoids from *E. officinalis* effectively reduced lipid levels in serum and tissues and had significant inhibitory effect on hepatic 3-hydroxy-3-methylglutaryl-CoenzymeA (HMG CoA) reductase activity [34].

Effect of amla on the lipid metabolism and protein expression involved in oxidative stress during the aging process were evaluated in laboratory rats. Sun Amla or ethyl acetate extract of amla, a polyphenol-rich fraction, on oral administration significantly increased the hepatic PPAR [α] protein level. Furthermore, the amla extracts reduced the expressions of hepatic NF-[kappa] B, inducible NO synthase (iNOS), and cyclo-oxygenase-2 (COX-2) protein levels which were increased with aging. The results suggested that amla may prevent age-related hyperlipidaemia through attenuating oxidative stress in the aging process [35].

E. officinalis (Amla), showed improvement in treatment of dyslipidemia and intima-media thickening and plaque formation in the aorta in hypercholesterolemic rabbits [36].

7. Clinical studies with *E. officinalis*

7.1 Hypolipidemic and anti-inflammatory activity

In a pilot clinical study the effect of *E. officinalis* extract (AMLAMAX ™) was evaluated on markers of systemic inflammation and dyslipidemia. Amlamax™ a purified, standardized, dried extract of amla containing about 35% galloellagic tannins along with other hydrolysable tannins showed reduction in total and LDL cholesterols, in blood CRP levels and enhancement of beneficial HDL cholesterol [37].

7.2 Hypoglycaemic activity

The hypoglycemic and lipid lowering effects of *E. officinalis* fruits were evaluated in normal and diabetic patients. The data showed a significant decrease ($p < 0.05$) in fasting and 2 hour post- prandial blood glucose levels along with total cholesterols (TC) and triglycerides (TG) in both normal and diabetic volunteers upon 21 days of treatment [38].

7.3 *W. somnifera*

W. somnifera (ashwagandha, WS) Family: Solanacae is widely used in Ayurvedic medicine, the traditional medical system of India and is an important medicinal plant, which is used in to cure many diseases. Some researchers have demonstrated that *W. somnifera* possesses powerful antioxidants. Preclinical studies also suggested the herb to produce an increase in the levels of natural antioxidants- superoxide dismutase, catalase and glutathione peroxidase [39].

8. Preclinical studies with *W. somnifera*

8.1 Anti-inflammatory and Antistress activities

Anti-inflammatory properties have been investigated to validate its use in inflammatory arthritis and animal stress studies have been performed to investigate its use as an antistress agent [40, 41].

8.2 Hypoglycaemic and Hypolipidemic activities

In a study flavonoids were isolated from the extracts of *W. somnifera* root and leaf and further hypoglycaemic and hypolipidemic effects were investigated in alloxan-induced diabetic rats. Eight weeks of treatment with *W. somnifera* and glibenclamide restored the changes in parameters to normal, indicating that it possesses hypoglycaemic and hypolipidemic activities [42].

8.3 Anti-oxidant activity

Researchers at Banaras Hindu University in Varanasi have discovered that some of the chemicals found in W. somnifera are powerful antioxidants. Studies conducted on rat brains showed that the herb increased the levels of superoxide dismutase, catalase, and glutathione peroxidase [39].

8.4 Anti-carcinogenic activity

The anti-carcinogenic property of Ashwagandga has been confirmed. Animal cell cultures has shown that the herb lowers [43] tumor size [44]. In another study, the herb was examined for its antitumor effects on urethane-induced lung tumors in adult male mice. After administration of ashwagandha for a period of seven months, the histological study of the lungs was similar to that observed in the lungs of control animals [45, 46].

9. Clinical studies with *W. somnifera*

9.1 Hypoglycemic and hypolipidemic activity

The hypoglycemic, hypocholesterolemic and diuretic effects of Ashwagandha has been studied in human clinical trials. A decrease in blood sugar levels comparable to that caused by the administration of a hypoglycaemic drug has been observed. Significant increases in urinary sodium, urine volume, and decreases in serum cholesterol, triglycerides, and low-density lipoproteins were also observed [46].

9.2 *O. sanctum*

O. sanctum also known as Tulsi belonging to family: Labiatae and its extracts are used in ayurvedic remedies. The use of this herb has been reported in the Indian traditional medical system, and its modern uses receive widespread attention over the years. Various parts of the plant have been claimed to be valuable in a wide range of diseases. It has been observed that Tulsi exerts hypocholesterolemic, hypotriglyceridemic and hypophospholipidemic effects. Among the chemical constituents contained in essential oil of *O. sanctum* leaves eugenol, a phenolic compound, is considered to be an active ingredient contributing for its hypolipidemic and antioxidant action [47].

9.3 Antioxidant and antineoplastic activity

In a study at Bangladesh, the antioxidant activity of Tulsi leaves extract was evaluated invitro. *O. sanctum* extract showed significant free radical scavenging activity. In the same study, antineoplastic activity of *O. sanctum* was demonstrated against Ehrlich Ascites Carcinoma (EAC) in mice. Tulsi leaves extract was administered at a dose of $50 mgKg^{-1}$ body weight intraperitoneally. Heamatological studies reveal that hemoglobin levels were reduced in EAC-treated mice, while near-normal recovery was observed in extract-treated animals. There was also a significant decrease in RBC count and an increase in WBC count in extract-treated mice compared to EAC-treated animals. From the results it was concluded that the extract has significant antioxidant and antineoplastic activity [48].

9.4 Hypolipidemic activity

In a study administration of fresh leaves of O. sanctum for four weeks resulted in significant changes in the lipid profile of normal albino rabbits. Significant reduction in serum total cholesterol, triglyceride, phospholipid and LDL cholesterol levels and an increase in stool HDL cholesterol and total sterol content were recorded [49].

Hypolipidemic activity of shade dried leaf powder of Tulsi along with the extracts and their fractions have shown *invitro* hypolipidemic and anti-peroxidative activity at very low concentrations in male albino rabbits. Aqueous extract feeding also provided significant protection of liver and aortic tissue from hypercholesterolemia-induced peroxidative damage [50].

10. Protection against radiation induced lipid peroxidation

A study was conducted to see if aqueous extract of *O. sanctum*, protects against radiation induced lipid peroxidation in liver and to determine the role, if any, of the inherent antioxidant system in producing radioprotection. Glutathione (GSH) and the antioxidant enzymes glutathione S-transferase (GST), reductase (GSRx), peroxidase (GSPx) and superoxide dismutase (SOD), as well as lipid peroxide (LPx) activity were estimated in the liver of adult swiss mice. The mice were injected intraperitoneally with 10 mg/kg of Tulsi for 5 consecutive days and exposed to 4.5 Gy of gamma radiation 30 min after the last injection. The aqueous extract itself increased GSH and enzymes significantly above normal levels, while irradiation significantly reduced all levels. The maximum drop was 30–60 min for GSH and related enzymes and 2 h for SOD. Pretreatment with the extract controlled the radiation-induced depletion of GSH and all enzymes and kept their levels within or above the control range. Irradiation significantly increased the lipid peroxidation rate, reaching a maximum value (about 3.5 times that of control) 2 hours after exposure. Aqueous extract pretreatment significantly reduced lipid peroxidation and accelerated recovery to normal levels [51].

11. Clinical studies with *O. sanctum*

In a study Tulsi leaves were tested on anthropometric measurements, diabetic symptoms and blood pressure in male patients with non-insulin dependent diabetes mellitus. Daily dosage of four capsules i.e. 2 g powder (Lunch and dinner) was given and supplementation was carryout for a period of 3 months. Significant percent reduction in the symptoms like polydypsia (35%), polyphagia (21%), and headache (27%), was observed in patients treated with Tulsi. It was concluded from the study that tulsi leaves are helpful in reducing subjects' diabetic symptoms and blood pressure. No significant improvement in subjects' anthropometric parameters was observed tulsi leaf powder supplementation [52].

11.1 *Boerhaavia diffusa*

Boerhaavia diffusa is a medicinal plant widely used in Ayurvedic medicine. The plant was named in the honor of Hermann Boerhaave, a famous Dutch physician of the 18th century [53]. It is also known as Spreading Hogweed in English, belonging to family, Nyctaginaceae.

12. Preclinical studies with *Boerhaavia diffusa*

12.1 Hypolipidemic activity

The efficacy of antioxidant and hypolipidemic agents tocotrienols and *Boerhaavia diffusa* by analyzing all the parameters in plasma lipoprotein lipids, total lipids (TL), total cholesterol (TC), triglycerides (TG), VLDL-C, LDL-C, HDL-C and MDA in oxidized cholesterol feeded rats. In the same study invitro oxidizability of LDL, was also demonstrated. All the plasma lipid parameters and MDA levels were significantly increased in hyperlipidemic control rats. After 4 weeks of administration of tocotrienols and *Boerhaavia diffusa* significantly reduced the overall oxidative burden and effectively ameliorated the above altered parameters. Thus indicating a strong hypolipidemic/antiatherogenic and antioxidant effect [54].

12.2 Anti-oxidant activity

The root extracts of *Boerhaavia diffusa* were evaluated using different solvents for free radical scavenging activity (FRSA) at a dose of 1000 mg/Kg body weight, prior to irradiation with 8 Gy gamma radiation as compared to mice pre-treated with extract at the dose of 250 and 500 mg/Kg body weight prior to irradiation with same dose of radiation. The data obtained showed that hydroethanolic extract produced potent free radical scavenging activity in DPPH., ABTS.+ and NO. assays and was found to be beneficial in reducing symptoms of radiation sickness, changes in body weight and mortality were minimum in the experimental animals. The antioxidant effect of *B. diffusa* roots was attributed to the presence of certain phenolic constituents like quercetin, caeffic acid, kempferol and their derivatives [55].

12.3 Hypoglycemic activity

Another study focused on blood glucose concentration and hepatic enzymes in normal and alloxan induced diabetic rats after daily oral administration of aqueous solution of *Boerhaavia diffusa* L. leaf extract (200 mg/kg) for 4 weeks. Significant improvement was recorded in blood glucose and glycosylated hemoglobin A1C levels. The action of hepatic enzymes such as hexokinase was significantly increased. Similarly, glucose-6-phosphatase, fructose-1,6-bisphosphatase were significantly decreased by the administration of BLEt in normal and diabetic rats. The results of BLEt were more potent when compared with antidiabetic drug—glibenclamide (600 μg/kg) [56].

12.4 Hepatoprotective activity

The hepato-protective activity of *Boerhaavia diffusa* alcoholic extract of the whole plant administered orally, was evaluated against experimentally induced hepatotoxicity using carbon tetrachloride in rats and mice. The extract also produced an increase in normal bile flow, indicating potent choleretic activity and no signs of toxicity were observed up to an oral dose of 2 g/kg.

13. Clinical studies with *Boerhaavia diffusa*

13.1 Antioxidant and Hypolipidemic activity

The effects of methanolic extract of Boerhaavia diffusa on oxidative stress in healthy and diabetes mellitus patients were studied. Results through this research demonstrated that diabetic patients experience increased oxidative stress when compared with normal subjects, significant increase in plasma, TG, TC, VLDL-C, LDL-C, and decrease in HDL-C. This may be due to markedly increased production of oxidant and significantly diminished antioxidant defense including a decline in total plasma antioxidant power. The study depicted that daily intake of *Boerhaavia diffusa* extract by diabetes mellitus patients significantly reduced TC, TG, LDL-C and increased HDL-C levels. The study concluded that extract of *Boerhaavia diffusa* may be useful in the prevention and treatment of the diabetes-induced hyperlipidemia and atherosclerosis. In addition, daily use of *Boerhaavia diffusa* can be efficacious and cost effective and good source of natural antioxidant [57].

14. Discussion

Over the past three decades, various experimental startegies have revealed the existence of cellular functions of peroxisomes related to reactive oxygen species (ROS) and reactive nitrogen species (RNS), and the function of peroxisomes as key centers of the cellular signaling apparatus. Peroxisomes of different origins have been detected which strongly indicate the interest of them as a cellular source of various signaling molecules, including ROS. In this review, we have focused on the generation and regulation of ROS in peroxisomes and the different antioxidant systems in this cell organelle [58]. We also enlighten the supporting evidence for application of antioxidants in preclinical and clinical evaluation of herbal supplements used in the management of associated disease complications.

Uncontrolled ROS production leads to structural modification of cellular proteins and alteration of their functions, resulting in cellular dysfunction and dysregulation of important cellular processes [59, 60]. Enhanced levels of ROS cause lipid, protein, and DNA damage. Specifically ROS can distort the lipid membrane and increase the fluidity and permeability of the membrane. Impairment of protein includes site-specific amino acid modification, peptide chain fragmentation, aggregation of crosslinked reaction products, modification of electric charges, immobilization of enzymes, and sensitivity to proteolysis [61]. Eventually, ROS can damage DNA by oxidizing deoxyribose, strand breakage, removal of nucleotides, changes in bases, and crosslinking DNA protein [62–65].

Literature suggests that peroxisomes are powerful and metabolically active organelles and are a very vital source of reactive oxygen species (ROS), H2O2, O2 (.-) and · OH, which are the products of diverse metabolic pathways, such as, photorespiration, fatty acid β-oxidation, nucleic acid and polyamine catabolism, ureide metabolism, to name a few. ROS were originally associated with oxygen toxicity. However, these reactive species also play a significant role in the signaling network that regulates essential processes in the cell. Peroxisomes have the ability to produce and scavenge H2O2 and O2(.-) rapidly, allowing to regulate dynamic alterations in ROS levels. The flexibility of these organelles, and based on varied developmental and environmental stimuli, render these organelles to perform a pivotal role in cellular signal transduction. The catalase and glycolate oxidase loss-of-function mutants have provided insights to study the consequences of modifications in endogenous H2O2 levels in peroxisomes. This has also facilitated the understanding of transcriptomic profile of genes regulated by peroxisomal ROS. It is now well established that peroxisomal ROS are involved in complicated signals which employ hormones, calcium, and redox homeostasis [66].

Antioxidants render an important role in these defense mechanisms. The antioxidat therapies target for maintenance of critical balance between oxidants and proxidants. In aerobic organisms, the steady release of free radicals needs to be equalized at the same degree of utilization of antioxidant. The naturally occuring enzymatic or non-enzymatic antioxidant systems prevent the formation of free radicals, and neutralize or repair the damage caused by them [62]. A wide range of endogenous and exogenous antioxidants are responsible, for providing protection against oxidative damage leading to development of chronic diseases [67]. The different types of antioxidant systems present both in plants [68] and the human body, contributes for controlling ROS homeostasis [69]. Release of natural ROS by the mitochondrial respiratory chain suggests that under certain conditions ROS can be metabolically beneficial but at the same time may also be harmful to cells [70–72].

The plant kingdom has served the mankind since ancient time and has provided remedies for various disease conditions. Over the period of time as the knowledge of plant derived medicines got advanced, it opened new avenues in improving the health and quality of life. Since many centuries, herbal drugs have been used both as food supplements and for medicinal requirements. When we mention about herbal medicine, it constitutes all parts of the plant like seeds, roots, bark, leaves, flowers and fruits from trees [73]. Most of the plant derived products and herbs act as potent scavengers of ROS or possess antioxidant activity. The phytoconstituents present in these herbs have been evaluated in numerous studies and are proved to rapidly stimulate the natural antioxidant enzyme systems such as catalase, superoxide dismutase, reduced glutathione etc., which protect the cells from oxidative damage and from progression of chronic diseases [74].

15. Conclusion

The conclusion of the present review is that peroxisomes are most common type of single layered membrane organelles identified in different types of eukaryotic cells. The origin of peroxisomes are through growth and division of cell and are independent organelles. These are also recognized as one of the most important and strong multifunctional organelles. The peroxisomes are able to facilitate the dynamic rotation of ROS generation and removal, fatty acid oxidation, β-oxidation of long-chain fatty acids, decomposition of purines, and glycerol, ether lipid and bile acid biosynthesis [75]. The metabolic processes of peroxisomes those which take place together with mitochondrial involvement are fatty acid β-oxidation and amino acid metabolism, but whereas the oxidation of different substrates is promoted by oxidases that consume oxygen. Several investigators in their work indicate the ability of peroxisomes to utilize 20% of the total oxygen consumption and can release up to 35% of the cellular hydrogen peroxide due to which are known to be major contributors of oxidative metabolism and in conserving oxidation balance [76]. Evidences from research show that the regulation of cell proliferation, apoptosis and carbohydrate metabolism is governed by hydrogen peroxide that act as a vital signaling molecule. However, at increased levels hydrogen peroxide is toxic and requires a regular check for its concentration. The other vital function of peroxisomes includes the action of antioxidant enzyme systems like the CAT, SOD, PRDX1, and PRDX5 [77–80]. CAT being the most significant enzyme and other antioxidant enzymes which metabolize the peroxidase hydrogen produced as a byproduct of peroxidases. Additionally super oxide dismutase 1 (SOD1) is regarded as a perfect peroxisomal protein, and a potent antioxidant enzyme that quenches the superoxide and accelerates the modification of oxygen to superoxide anion ($O2^-$) [81].

Plant-based bioactive molecules have received a lot of recognition since the past few decades. Several studies have demonstrated their therapeutic significance in the management of disease conditions and for prevention as well. The complete phytochemical profile of whole plants, plant extracts or even the isolated constituents are well explained in the literature, which can be utilized for planning treatment strategies for various diseases including diabetes mellitus, cardiovascular disease and neurodegenerative disorders. Extensive randomized trials are warranted to collect data for establishing the medical interest or probable hazards of antioxidant supplementation.

There is enormous substantiation that oxidative stress has been implicated in normal physiological processes and environmental interactions that occur in a cell. Several mechanisms are involved in antioxidant defense systems that render protection against oxidative damage. Literature suggests that in many conditions these processes seem to be tangled. ROS profusely disrupts the antioxidant balance, causing oxidative stress and results in constant alterations in the cellular material, which includes carbohydrate, protein and lipid substances [70, 82–84]. It is likely to presume that oxidative stress can be a cause of tissue damage and finally arresting the natural cellular-signaling processes. However, a thorough understanding of the biochemical events occurring at a cellular level to influence oxidative damage is mandatory to direct ensuing progress. Peroxisomes serve as very important sites for detoxification of ROS. But however peroxisomes itself release these radicals. With the advent of fluorescence methods and having vast knowledge of peroxisomal functions, we expect to read more about the role of this organelle in the near future that can be useful in the treatment of related disorders.

Acknowledgements

I thank Al Hawash Private University, Homs, Arab Republic of Syria, for providing the necessary support in completion of this chapter.

Author details

Nishat Fatima
Al Hawash Private University, Al-Mouzeina, Syria

*Address all correspondence to: nishat_fatima50004@yahoo.com

References

[1] Knight JA. Review: Free radicals, antioxidants and immune system. Annals of Clinical and Laboratory Science. 2000;**30**:145-158

[2] Seis H. Biochemistry of oxidative stress. Angewandte Chemie International Edition. 1986;**25**:1058-1071

[3] Lum H, Roebuck AK. Oxidant stress and endothelial cell dysfunction. American Journal of Physiology. Cell Physiology. 2001;**280**:C719-C741

[4] Cooper GM. The Cell: A Molecular Approach. 2nd ed. Sunderland (MA): Sinauer Associates; 2000

[5] Seis H, editor. Antioxidants in Disease, Mechanism and Therapy. New York: Academic Press; 1996

[6] Koppenol WH, Moreno JJ, Pryor WA, Ischiropoulos H, Beckman JS. Peroxynitrite, a cloaked oxidant formed by nitric oxide and super oxide. Chemical Research in Toxicology. 1992;**5**:834-842

[7] Griendling KK, Fitz Gerald GA. Oxidative stress and cardiovascular injury. Part I: Basic mechanisms and invivo monitoring of ROS. Circulation. 2003;**108**:1912-1916

[8] Milstien S, Katusic Z. Oxidative of tetrahydrobiopterin by peroxynitrite. Implications of vascular endothelial function. Biochemical and Biophysical Research Communications. 1999;**263**: 681-684

[9] Landmesser U, Speakermann S, Dikalov S, Tatge H, Wilke R, Kohler C, et al. Vascular oxidative stress and endothelial dysfunction in patients with chronic heart failure: Role of xanthine-oxidse and extracellular superoxide dismutase. Circulation. 2002;**106**:3073-3078

[10] Du X, Matsumura T, Edelstein D, Rossetti L, Zsengeller Z, Szabo C, et al. Inhibition of GAPDH activity by poly (ADP-ribose) polymerase activates three major pathways of hyperglycemic damage in endothelial cells. The Journal of Clinical Investigation. 2003;**112**:1049-1057

[11] Packer L, Ong ASH, editors. Biological Oxidants and Antioxidants: Molecular Mechanisms and Health Effects. Champaign: AOCS Press; 1998

[12] Devasagayam TPA, Tilak JC, Boloor KK, Ketaki SS, Ghaskadbi SS, Lele RD. Free radicals and antioxidants in human health: Current status and future prospects. Journal of the Association of Physicians of India. 2004;**52**:794-804

[13] Upritchard JE, Sutherland WH, Mann JI. Effect of supplementation with tomato juice vitamin E and vitamin C on LDL oxidation and products of inflammatory activity in type 2 diabetes. Diabetes Care. 2000;**23**:733-738

[14] Reaven PD, Herold DA, Barnett J, Edelman S. Effects of vitamin E on susceptibility of low density lipoprotein and low density lipoprotein subfractions to antioxidation and on protein glycation in NIDDM. Diabetes Care. 1995;**18**:807-816

[15] Pinkney JH, Downs L, Hopton M, Mackness MI, Bolton CH. Endothelial dysfunction in type 1 diabetes mellitus: Relationship with LDL oxidation and the effects of vitamin E. Diabetes Medications. 1999;**16**:993-999

[16] Cunningham JJ, Ellis SL, McVeiigh KL, Levine RE,

Calles-Escandon J. Reduced mononuclear leukocyte ascorbic acid content in adults with insulin-dependent diabetes mellitus consuming adequate dietary vitamin C. Metabolism. 1991;**40**:146-149

[17] Ting HH, Timmi FK, Boles KS, Creager SJ, Gnaz P, Creager MA. Vitamin C improves endothelium dependent vasodilatation in patients with non-insulin dependent diabetes mellitus. The Journal of Clinical Investigation. 1996;**97**(1):22-28

[18] Vlachopoulos C, Alexopoulos N, Stefanadis C. Effect of dark chocolate on arterial function in healthy individuals: Cocoa instead of ambrosia? Current Hypertension Reports. 2006;**8**:205-211

[19] Allison P. Global Survey of Marine and Estuarine Species Used for Traditional Medicine and Tonic Foods. WHO Report. Quebee, Canada: McGill University; 1996

[20] Anonymous Medicinal Plants; their Biodiversity, Screening and Evaluation. New Delhi: Center for Science and Technology of the Non-Aligned and Other Developing Countries; 1998

[21] Bhattacharjee SK. Handbook of Meidicinal Plants. Jaipur, India: Pointer Publishers; 1998

[22] Veena S, Sadhna P, Ritu P. Withania somnifera: A rejuvenating ayurvedic medicinal herb for the treament of various human ailments. International Journal of Pharm Tech Research. 2011;**3**(1):187-192

[23] Verma N, Vinayak M. Effect of Terminalia arjuna on antioxidant defense system in cancer. Molecular Biology Reports. 2009;**36**(1):159-164

[24] Sumitra M, Manikandan P, Kumar AD, Arutselven N, Balakrishna K, Manohar MB, et al. Experimental myocardial necrosis in rats: Role of arjunolic acid on platelet aggregation, coagulation and antioxidant status. Molecular and Cellular Biochemistry. 2001;**224**:135-142

[25] Alam MS, Kaur G, Ali A, Hamid H, Ali M, Athar M. Two new bioactive oleanane triterpene glycosides from Terminalia arjuna. Natural Product Research. 2008;**22**(!4):1279-1288

[26] Bharani A, Ahirwar LK, Jain N. Terminalia Arjuna reverses impaired endothelial function in chronic smokers. Indian Heart Journal. 2004;**56**(2):123-128

[27] Bharani A, Ganguli A, Mathur LK, Jamra Y, Raman PG. Efficacy of Terminalia arjuna in chronic stable angina: A double-blind, placebo-controlled, crossover study comparing Terminalia arjuna with isosorbide mononitrate. Indian Heart Journal. 2002;**54**(2):170-175

[28] Satyavati GV, Raina MK, Sharma M. Medicinal Plants of India. Vol. I. New Delhi: Indian Council of Medical Research; 1976. pp. 377-379

[29] Thangaraj R, Rathakrishna SA, Pannersevlam M, Jayaraman B. Antioxidant property of Emblica oficanalis during experimentally induced restrain stress in rats. Journal of Health Sciences. 2007;**53**(4):496-499

[30] Ghoshal S, Trripathi VK, Chauhan S. Active constituents of Emblica Officianalis. Part I. the chemistry and antioxidant effects of two new hydrolysable tannins. Emblicanin a and B. Indian Journal of Chemistry. 1996;**35**:941-948

[31] Qinna NA, Kamona BS, Alhussainy TM, Taha H, Baddwan AA, Matalka KZ. Effect of prickly pear dried

leaves, artichoke leaves, turmeric and garlic extracts, and their combinations on preventing Dyslipdemia in rats. International Scholarly Research Network ISRN Pharmacology. **2012**:167979

[32] Mishra M, Pathak UN, Khan AB. Embilica officianalis Gaertn and serum cholesterol level in experimental rabbits. British Journal of Experimental Pathology. 1981;**62**(5):526-528

[33] Anila L, Vijaylakshmi NR. Flavonoids from Emblica officianalis and Mangifera Indica –effectiveness for dyslipidemia. Journal of Ethnopharmacology. 2002;**79**:81-87

[34] Yokozawa T, Kim HY, Kim HJ, Okubo T, Chu DC, Juneja LR. Amla (Emblica officinalis Gaertn.) prevents dyslipidemia and oxidative stress in the ageing process. The British Journal of Nutrition. 2007;**97**(6):1187-1195

[35] Antony B, Merina B, Sheeba V, Mukkadan J. Effect of standardized Amla extract on atherosclerosis and dyslipidemia. Indian Journal of Pharmaceutical Sciences. 2006;**68**(4):437-441

[36] Antony B, Benny M, Kaimal TNB. Apilot clinical study to evaluate the effect of Embilica offianalis extract (AMLAMAX™) on markers of systemic inflammation and dyslipidemia. Indian Journal of Clinical Biochemistry. 2008;**23**(4):378-381

[37] Muhammad SA, Ayesha R, Amanat A, Ahmad M. Effect of Amla fruit (Emblica officianalis) on blood glucose and lipid profile of normal subjects and type 2 diabetic patients. International Journal of Food Sciences and Nutrition. 2011;**62**(6):606-616

[38] Dhuley JN. Adaptogenic and cardioprotective action of ashwagandha in rats and frogs. Journal of Ethnopharmacology. 2000;**70**:57-63

[39] Somasundaram S, Sadique J, Subramoniam A. Influence of extra-intestinal inflammation on the in vitro absorption of 14C-glucose and the effects of anti-inflammatory drugs in the jejunum of rats. Clinical and Experimental Pharmacology & Physiology. 1983;**10**:147-152

[40] Dhuley JN. Effect of ashwagandha on lipid peroxidation in stress induced animla. Journal of Ethnopharmacology. 1998;**60**:173-178

[41] Udayakumar R, Kasthurirengan S, Salammal MT, Rajesh M, et al. Hypoglycaemic and Hypolipidaemic effects of Withania somnifera root and leaf extracts on Alloxan-induced diabetic rats. International Journal of Molecular Sciences. 2009;**10**:2367-2382

[42] Ichikawa H, Takada Y, Shishodia S, Jayaprakasam B, Nair MG, Aggarwal BB. Withanolides potentiate apoptosis, inhibit invasion, and abolish osteoclastogenesis through suppression of nuclear factor-kappaB (NF-kappaB) activation and NF-kappaB-regulated gene expression. Molecular Cancer Therapeutics. 2006;**6**:1434-1445

[43] Prakash J, Gupta SK, Dinda AK. Withania somnifera root extract prevents DMBA induced squamous cell carcinoma of skin swiss albino mice. Nutrition and Cancer. 2002;**42**:91-97

[44] Singh N, Singh SP, Nath R, et al. Prevention of urethane induced lung adenomas by Wihania somnifera (L) dunal in albino mice. International Journal Crude Drug Research. 1986;**24**:90-100

[45] Andallu B, Radhika B. Hypoglycemic, diuretic and hypocholesterolemic effect

of inter cherry (Withania somnifera) root. Indian Journal of Experimental Biology. 2000;**3**:607-609

[46] Thamolwas S, Watcharapora DNA, Thanapat S, Thirawarapan S, Poungshompoo S. Antioxidant activity and lipid-lowering effect of essential oils extracted from Ocimum sanctum leaves in rats fed with a high cholesterol diet. Journal of Clinical Biochemistry and Nutrition. 2010;**46**(1):52-59

[47] Islam MS, Alam BM, Zahan R, Sarker GC, et al. Invitro antioxidant and antineoplastic activities of Ocimum santum in Ehrlich ascites carcinoma bearing mice. International Journal of Cancer Research. 2011;**7**(3):209-221

[48] Sarkar A, Lavania C, Pandey DN, Pant MC. Changes in the blood lipid profile after administration of Ocimum sanctum (Tulsi) leaves in the normal albino rabbits. Indian Journal of Physiology and Pharmacology. 1994;**38**(4):311-312

[49] Geetha RK, Vasudevan DM. Inhibition of lipid peroxidation by botanical extracts of Ocimum sanctum and its role in protection against radiation injury. Life Sciences. 2004;**76**:21

[50] Umadevi P, Gansoundari A. Modulation of glutathione and antioxidant enzymes by Ocimum sanctum and its role in protein against radiation injury. Indian Journal of Experimental Biology. 1999;**37**:262

[51] Kochhar A, Sharma N, Rajbir SR. Effect of supplementation of Tulsi (Ocimum sanctum) and neem (Azadirachta indica) leaf powder on diabetic symptoms, anthropometric parameters and blood pressure of non-insulin dependent male diabetics. Ethno-Medicine. 2009;**3**(1):5-9

[52] Lad V. The Complete Book of Ayurvedic Home Remedies. New York: Three Rivers Press; 1999

[53] Amir K, Abhay SC, Fouzia I, Chettri S, Malhotra D. Therapeutic impacts of tocotrienols and Boerhaavia diffusa on cholesterol dynamics, lipid hydroperoxidation and antioxidant status on hyperlipidemic rats: Induced by oxidized cholesterol. Recent Research in Science and Technology. 2011;**3**(11):13-21

[54] Dharkar P, Saini MR, Sharma J. Comparative evaluation of free radical scavenging activity of Boerhaavia diffusa root extracts (BDRE) and determination of dose effectivity against radiation induced damages in Swiss albino mice. International Journal of Applied Research in Natural Products. 2012;**5**(4):9-18

[55] Pari L, Satheesh MA. Antidiabetic activity of Boerhaavia diffusa L. effect on hepatic key enzymes in experimental diabetes. Journal of Ethnopharmacology. 2004;**91**(1):109-113

[56] Chandan BK, Sharma AK, Anand KK. Boerhaavia diffusa: A study of its hepatoprotective activity. Journal of Ethnopharmacology. 1991;**31**(3):299-230

[57] Amir K, Malhotra D, Abhay SC, Chettri S, Fouzia I, Rathore NS. Invitro antioxidant effect of Boerhaavia diffusa on copper mediated oxidative modification of LDL in type-II diabetic pateints. International Journal of Pharmaceutical Sciences. 2011;**2**(4):75-88

[58] del Río LA, López-Huertas E. ROS generation in peroxisomes and its role in cell Signaling. Plant & Cell Physiology. 2016;**57**(7):1364-1376. DOI: 10.1093/pcp/pcw076

[59] Finkel T, Holbrook NJ. Oxidants, oxidative stress and the biology of

ageing. Nature. 2000;**200**(408):239-247. DOI: 10.1038/35041687

[60] Kaminski KA, Bonda TA, Korecki J, Musial WJ. Oxidative stress and neutrophil activation—The two keystones of ischemia/reperfusion injury. International Journal of Cardiology. 2002;**86**:41-59. DOI: 10.1016/s0167-5273(02)00189-4

[61] Ayala A, Munoz MF, Arguelles S. Lipid peroxidation: Production, metabolism, and signaling mechanisms of malondialdehyde and 4-hydroxy-2-nonenal. Oxidative Medicine and Cellular Longevity. 2014;**2014**:360438

[62] Sharma P, Jha AB, Dubey RS, Pessarakli M. Reactive oxygen species, oxidative damage, and antioxidative defense mechanism in plants under stressful conditions. Journal of Botany. 2012;**2012**:26

[63] Cadet J, Wagner JR. DNA base damage by reactive oxygen species, oxidizing agents, and UV radiation. Cold Spring Harbor Perspectives in Biology. 2013;**5**:a012559. DOI: 10.1101/cshperspect.a012559

[64] Cadet J, Davies KJA, Medeiros MHG, Di Mascio P, Wagner JR. Formation and repair of oxidatively generated damage in cellular DNA. Free Radical Biology & Medicine. 2017;**107**:13-34. DOI: 10.1016/j.freeradbiomed.2016.12.049

[65] Liang X, Wang S, Wang L, Ceylan AF, Ren J, Zhang Y. Mitophagy inhibitor liensinine suppresses doxorubicin-induced cardiotoxicity through inhibition of drp1-mediated maladaptive mitochondrial fission. Pharmacological Research. 2020;**157**:104846. DOI: 10.1016/j.phrs.2020.104846

[66] Luisa MS, María RS, María CRP, Luis ADR. Role of peroxisomes as a source of reactive oxygen species (ROS) Signaling molecules. Sub-Cellular Biochemistry. 2013;**69**:231-255. DOI: 10.1007/978-94-007-6889-5_13

[67] Clark IA, Cowden WB, Hunt NH. Free radical-induced pathology. Medicinal Research Reviews. 1985;**5**:297-332

[68] Cadet J, Ravanat JL, Tavernaporro M, Menoni H, Angelov D. Oxidatively generated complex DNA damage: Tandem and clustered lesions. Cancer Letters. 2012;**327**:5-15. DOI: 10.1016/j.canlet.2012.04.005

[69] Hsu TC, Young MR, Cmarik J, Colburn NH. Activator protein 1 (AP-1)- and nuclear factor kappaB (NF-kappaB)-dependent transcriptional events in carcinogenesis. Free Radical Biology & Medicine. 2000;**28**:1338-1348. DOI: 10.1016/s0891-5849(00)00220-3

[70] Birben E, Sahiner UM, Sackesen C, Erzurum S, Kalayci O. Oxidative stress and antioxidant defense. World Allergy Organization Journal. 2012;**5**:9-19

[71] Poli G, Leonarduzzi G, Biasi F, Chiarpotto E. Oxidative stress and cell signalling. Current Medicinal Chemistry. 2004;**11**:1163-1182. DOI: 10.2174/0929867043365323

[72] Valko M, Leibfritz D, Moncol J, Cronin MT, Mazur M, Telser J. Free radicals and antioxidants in normal physiological functions and human disease. The International Journal of Biochemistry & Cell Biology. 2007;**39**:44-84. DOI: 10.1016/j.biocel.2006.07.001

[73] Egan CD. Addressing the use of herbal medicine in the primary care setting. Journal of the American Association of Nurse Practitioners. 2002;**14**:166-171

[74] Sharifi-Rad M, Anil Kumar NV, Zucca P, Varoni EM, Dini L, Panzarini E, et al. Lifestyle, oxidative stress, and antioxidants: Back and forth in the pathophysiology of chronic diseases. Frontiers in Physiology. 2020;**11**:694

[75] Wanders R. Biochemistry of mamma-lian peroxisomes revisited. Annual Review of Biochemistry. 2006;**75**:295-332

[76] Alberto B, Nozomu O, Britton C. The cellular production of hydrogen peroxide. The Biochemical Journal. 1972;**128**:617-630. DOI: 10.1042/bj1280617

[77] Schrader M, Yoon Y. Mitochondria and peroxisomes: Are the 'big brother' and the 'little sister' closer than assumed? BioEssays. 2010;**29**:1105-1114

[78] Bonekamp NA, Völkl A, Fahimi HD, Schrader M. Reactive oxygen species and peroxisomes: Struggling for balance. BioFactors (Oxford, England). 2010;**35**:345-355

[79] Stolz DB, Zamora R, Vodovotz Y, Loughran PA, Billiar TR, Kim YM, et al. Peroxisomal localization of inducible nitric oxide synthase in hepatocytes. Hepatology. 2002;**36**:81-89. DOI: 10.1053/jhep.2002.33716

[80] Rhee SG, Kang SW, Jeong W, Chang TS, Yang KS, Woo HA. Intracellular messenger function of hydrogen peroxide and its regulation by peroxiredoxins. Current Opinion in Cell Biology. 2005;**17**:183-189

[81] Dhaunsi GS, Gulati S, Singh AK, Orak JK, Singh I. Demonstration of Cu-Zn superoxide dismutase in rat liver peroxisomes. Biochemical and immunochemical evidence. Journal of Biological Chemistry. 1992;**267**:6870-6873

[82] Zal F, Taheri R, Khademi F, Keshavarz E, Rajabi S, Mostafavi-Pour Z. The combined effect of furosemide and propranolol on GSH homeostasis in ACHN renal cells. Toxicology Mechanisms and Methods. 2014;**24**:412-416. DOI: 10.3109/15376516.2014.926437

[83] Salehi B, Martorell M, Arbiser JL, Sureda A, Martins N, Maurya PK, et al. Antioxidants: Positive or negative actors? Biomolecules. 2018;**8**:124. DOI: 10.3390/biom8040124

[84] Sharifi-Rad J, Sharifi-Rad M, Salehi B, Iriti M, Roointan A, Mnayer D, et al. In vitro and in vivo assessment of free radical scavenging and antioxidant activities of *Veronica persica* Poir. Cellular and Molecular Biology. 2018;**64**:57-64

Chapter 5

Peroxisomal Modulation as Therapeutic Alternative for Tackling Multiple Cancers

Shazia Usmani, Shadma Wahab, Abdul Hafeez, Shabana Khatoon and Syed Misbahul Hasan

Abstract

Peroxisomes are indispensably involved as a central player in the metabolism of reactive oxygen species, bile acids, ether phospholipids, very-long-chain, and branched-chain fatty acids. The three subtypes of PPARs are PPAR-alpha, PPAR-delta, and PPAR-gamma which have been found to be instrumental in the control of cancer metabolism cascades. Any disproportionate expression of PPAR can lead to the progression of cell growth and survival in diverse types of cancers. It can be exploited both as an agonist or antagonist for utilization as a potential therapeutic alternative for the treatment of cancer. Therefore, the multifunctional PPAR modulators have substantial promise in various types of cancer therapies. Many recent studies led to the observations that a variety of phytochemicals, including phenolics, have been implicated in anticancer effects. Plant phenolics seem to have both palliative and treatment opportunities in combating cancer which requires deep insight into the proposed mechanisms. Henceforth, this chapter highlights the role of peroxisomal subtypes as an activator or suppressor followed by its modulation through bioactive obtained from a variety of crude drugs. A discussion on various challenges restricting proper utilization has also been incorporated.

Keywords: peroxisome, metabolism, PPARs, herbal, cancer

1. Introduction

Peroxisomes are small membrane-bound organelles with simple structures but contain enzymes that display a wide range of metabolic activities. About 50 peroxisomal enzymes have been identified where major [1, 2] pathways for metabolism involve α- and β-oxidation of fatty acids, biosynthesis of ether lipids, polyamines, D-amino acids, glyoxylate, and purines. The synthesis and assembly of peroxisomal proteins occur on free ribosomes which are then imported into these tiny organelles as completed polypeptide chains. The disorders related to peroxisomal functions can be attributed to a disturbance in the formation of the organelles or might be related to defects in either a particular peroxisomal enzyme or a related transporter [3, 4]. The metabolic disorders promote the accumulation of substrates that are usually degraded by specific peroxisomal enzymes. A variety of clinical symptoms has proven to be very severe leading to an early death.

2. Metabolic implications of peroxisomes, a druggable target

Peroxisome-related homeostatic balance is an indispensable mechanism of health where the removal of worn-out and defective peroxisomes occurs through autophagy. Association with mitochondria is reflected when commotion of peroxisomal function results in disruption of mitochondrial function. The impaired peroxisomal function has been found to be instrumental in special conditions of neurodegenerative disorders and diabetes, while dysregulation in peroxisomal function can result in cancer [5, 6]. There has been increasing evidence linking peroxisomal misregulation to the eruption of several diseases which potentiate an elevated possibility of targeting peroxisomal involvement in disease prevention or treatment.

Peroxisomes are amazingly active organelles, which have an important role in lipid and hydrogen peroxide metabolism making them elemental for human health [7]. Despite great advances in identification of essential components and related molecular mechanisms, an understanding of the process by which peroxisomes are incorporated into metabolic pathways is of elementary importance. The interaction of peroxisomes with other subcellular compartments, metabolic co-operations, peroxisome–peroxisome interactions, and the interaction of peroxisomes with microtubules needs to be addressed to utilize this information directly to combat the process of disease development.

Peroxisomes are consigned to clearing up the reactive oxygen chemical debris cast off by other organelles, where their functions extend far beyond hydrogen peroxide metabolism [8]. Peroxisomes are closely associated with mitochondria, and their ability to carry out fatty acid oxidation and lipid synthesis may be highly implicated in generating cellular signals required for normal physiology. The biology of peroxisomes and their relevance to human disorders, including cancer, obesity-related diabetes, and degenerative neurologic disease cannot be undermined [9].

Peroxisomes are multifarious where they invariably modulate the metabolism of reactive oxygen species and primary homeostatic mechanisms, such as oxidation of fatty acid, synthesis of bile acid, and transport of cholesterol. Henceforth, it is implicative that peroxisomal homeostasis is an important regulator of health, and disruption of peroxisomal function can lead to mitochondrial dysfunction, reflecting the intimate link between the two organelles [10].

The impaired peroxisomal function leads to neurodegenerative disorders and diabetes, but dysregulation may have far-reaching effects, such as the development of cancer [11]. The peroxisomal function is also transformed with aging owing to deviations in the expression and/or localization of peroxisomal matrix proteins.

The homeostatic mechanisms of peroxisomes are undermined by the existence of distressing genetic disorders attributed to impaired peroxisomal function. However, with amplified evidence connecting peroxisomal dysfunction to the pathogenesis of these acquired diseases, it can be utilized as a druggable target in disease prevention or treatment [12].

The immune system evasion is one of the mainstays of cancer, and peroxisomes have an indispensable role in the regulation of cellular immune responses.

Investigations of individual peroxisome proteins and metabolites provide for their pro-tumorigenic functions [13]. It is, therefore, important to highlight new advances in our understanding of biogenesis, enzymatic functions, and autophagic degradation of peroxisomes, which shall avail enough evidence to link such activities to tumor development. Such findings shall add to the possibility of exploitation of peroxisome-related processes for efficient battling against cancer.

3. Peroxisome proliferator-activated receptors (PPARs)

With the above, emerging evidence, exploring the possible sites of activation of peroxisomal receptors could be intriguing with respect to the benefit and risk ratio.

In this context, it was deduced that activation of peroxisome proliferator-activated receptors (PPARs) can be considered an efficient strategy for the treatment of metabolic dysregulation [14]. An ample of new moieties having the prospects to stimulate peroxisome proliferation have been discovered in the recent past.

The receptor which was cloned from a mouse liver, and titled a peroxisome proliferator-activated receptor (PPAR) could regulate the expression of sizable genes involved in the regulation of glucose and lipid metabolism [15].

Besides, the ligands which activate PPARs lead to the promotion of co-activators and inhibition of co-repressors remodeling the chromatin and initiating transcription [16].

4. Metabolic regulation by PPARs and their repercussions

The peroxisome proliferator-activated receptors (PPARs) are a set of nuclear receptors namely PPAR gamma, PPAR alpha, and PPAR delta, encrypted by diverse genes. PPARs are ligand-regulated transcription factors regulating gene expression by binding to specific response elements (PPREs) within promoters. PPARs bind as heterodimers with a retinoid X receptor and, upon binding agonist, interact with cofactors such that the rate of transcription initiation is increased [17].

The PPARs are major regulators of lipid metabolism where fatty acids and eicosanoids have been recognized as common ligands. Synthetic PPAR ligands, such as fibrates and thiazolidinediones, have been effectively used in the treatment of dysregulation of lipids and glucose metabolism.

The discovery of these ligands led to the disclosure of many impending functions for the PPARs in pathological metabolic situations, such as demyelination, atherosclerosis, and cancer [18].

4.1 Peroxisome proliferator-activated receptor-alpha (PPAR-α)

It has been recognized as the nuclear receptor for a class of rodent hepato-carcinogens leading to the proliferation of peroxisomes. PPAR-α is a transcription factor that happens to be the major regulator of lipid metabolism in the liver [19].

It is primarily activated via ligand binding where fatty acids, such as arachidonic acid and their metabolites from the ligand groups. Another category consists of synthetic ligands, such fibrate drugs referred to as peroxisome proliferators [20].

4.2 Peroxisome proliferator-activated receptor beta or delta (PPAR-β or PPAR-δ)

PPAR- δ is a nuclear hormone receptor that manages diverse biological processes involved in the progression of several chronic ailments, viz. obesity, atherosclerosis, and cancer [21].

PPAR-δ act as an integrated unit for transcription regulation and nuclear receptor signaling. It stimulates the transcription of a wide variety of target genes by binding to specific DNA elements.

Many fatty acids and their derivatives induce PPAR δ viz. arachidonic acid and its metabolites [22].

4.3 Peroxisome proliferator-activated receptor gamma (PPAR-γ)

PPAR-γ or the glitazone reverse insulin resistance receptor, is a type II nuclear receptor that is encoded by the PPAR-γ gene in humans [23, 24]. The protein encoded by this gene is PPAR-γ, which regulates the differentiation of adipose cells [25]

When the activity of PPAR-γ is regulated via phosphorylation through the MEK/ERK pathway, it results in decreasing transcriptional activity of PPAR-γ. The result is a loss of insulin sensitivity due to diabetic gene modifications. Owing to the above reasons, PPAR-γ has been implicated in the pathology of numerous diseases, including obesity, diabetes, atherosclerosis, and cancer [26].

PPAR-γ controls fatty acid storage and metabolism of glucose. The genes activated by PPAR-γ stimulate lipid uptake and adipogenesis by fat cells. The agonists have been reported to be used in the treatment of hyperlipidemia and hyperglycemia [27].

PPAR-γ decreases the inflammatory response of many cardiovascular cells. PPAR-γ activates the paraoxonase-1 gene, resulting in an increase of paraoxonase 1 in the liver, which reduces the incidence of atherosclerosis [28].

The prevalence of metabolic syndromes is growing in the adult and pediatric groups which include majorly atherogenic dyslipidemia raised blood pressure, and pre-eminent plasma glucose [29].

Peroxisome proliferator-activated receptors (PPARs) may come up as potential therapeutic targets for the treatment or prevention of metabolic syndromes. Further, there is substantial evidence that its agonists are, therefore, used in the treatment of metabolic syndrome and cardiovascular diseases [30].

Activation of peroxisome proliferators-activated receptor (PPAR) is invariably indulged in varied mechanisms related to lipid profile.

One of the researches in this area confirmed the role of herbs in the stimulation of PPARα. Among the tested plant extracts, about nine had shown moderate PPARα transactivation [31]. The bioactive, piperine, and capsaicin revealed substantial trans-activational activities followed by a moderate activity in chalcones. It was concluded that a diet rich in natural products viz. herbs, act as PPARα agonists improving the lipid profile.

5. Proposed mechanisms of PPARs in tumor suppression

5.1 Distressing metabolism

PPAR ligands disturb the survival of cancer cells in such a way that the metabolism enters into complete devastation. Owing to the potential of PPAR ligands, they are been considered a potential source of anticancer agents, with minimal toxicities [32].

PPAR activation disrupts the metabolism of cancer cells mainly by blocking the synthesis of fatty acids and promoting fatty acid oxidation. Owing to nutrient depletion, in the tumor microenvironment, PPAR coordinates with AMP-dependent protein kinase in repressing oncogenic Akt activity, inhibiting cell proliferation, and inducing glycolysis-dependent cancer cells into "metabolic failure" [33].

There is substantial evidence for the antiproliferative role, and prevention of metastatic indulged by PPAR ligands, which prompts a detailed compilation on the possible potential of PPAR in tumor suppression [34].

5.2 PPAR subdues cell proliferation by overpowering inflammation

Suppression of inflammation is another mode contributing to anticancer effects. PPAR takeover the inflammation and activation of uncoupling proteins, which wanes the mitochondrial ROS generation and resultant cell proliferation. PPAR ligands can be considered as a low-toxic and well-tolerated therapeutic moiety to combat cancer [35].

The peroxisome proliferator-activated receptor γ ligands exhibited anticancer activity *in vitro*, against diverse neoplastic cells *whereas* animal studies also reflected that they are *in vivo* anticancer effects and chemopreventive proficiency. The effect may be attributed to slowing down the growth and induction of partial differentiation of several cancer cells, such as lipo-sarcoma, and cancers, such as colon, prostate, and breast cancers [36].

At the molecular level, these can decrease the levels of cyclin D_1 and E, nuclear factor κB, and inflammatory cytokines. Some relevant data support the fact that PPAR γ might act as a gene for tumor suppression. On the other hand, several captivating pieces of evidence, suggest that under certain specific settings, PPAR γ ligands can lead to cancer [37].

Yet, the bulk of studies still reflects the fact that PPAR γ ligands bear antiproliferative potential against numerous transformed cells and may be applied in adjuvant treatments strategies for several common tumors [37–39].

As per research by Morinishi et al., activation of PPAR-*α* seems to be involved in the control of colorectal carcinomas, where nuclear expression of PPAR-*α* may be established as an indispensable therapeutic target for the respective treatment of the disease. It was deduced that the nuclear expression of PPAR-*α* was significantly higher in subtly differentiated adenocarcinoma than in mucinous adenocarcinoma [40].

Colorectal cancer poses one major threat due to excessive dietary fat posing as a major threat. As it is involved in the regulation of lipid and carbohydrate metabolism, it needs to be studied extensively in this case [41]. Despite the fact, that researchers have scrutinized the expression and clinical repercussions of PPARs in colorectal cancer, the exact mechanism needs to be further explored.

Diverse studies have been undertaken, focusing on the assumed link between the polymorphisms and mutations of the PPAR *γ* gene with the incidence of cancer [42, 43].

Ikezoe et al. [44] analyzed 397 clinical samples and cell lines, including colon, breast, and lung cancers for mutations of the PPAR *γ* gene. They indicated the absence of PPAR *γ* gene mutations in the tested cell lines ascertaining PPAR *γ* mutations may occur in cancers but very rarely.

There has been substantial experimental data supporting that synthetic PPAR *γ* ligands induce apoptosis in several types of cancer cells [45, 46]. Albeit, the majority of the evidence has documented that PPAR *γ* agonists inhibit growth in cancer cells but the mechanism of the growth inhibition by PPAR γ agonists is not well understood and complicated.

5.3 Differential behavior of peroxisome

Specific tumors behave variably in terms of peroxisomal activity. It has been observed, thus, that the enzymatic activities of peroxisomal metabolism decline in the breast [47], colon [48], and hepatocellular carcinomas [49]. Similar observations were recorded in renal cell carcinoma [50]. In a related finding, von Hippel-Lindau (VHL)-deficient clear cell renal carcinoma displayed reduced peroxisomal activity. In contrast to this, some reports reveal that peroxisomal metabolic activities lead to enhancing the growth of tumors [51]. Few cancer cells count on peroxisomal lipid metabolism for energy and support the survival of cancer cells in the tumor micro-environment [52]. This controversial behavior of peroxisome indicates the fact that under a certain specific environment, it promotes or diminishes cancer growth, which may be attributed to the type of tumor. In this regard, it is implicative to further investigate the inducing factors that decide the fate of the metabolism of peroxisomes, closely related to its cancer proliferation effects.

Many studies have been undertaken to study the potential of the combinatorial approach where PPAR agonists can be used for the treatment of resistant cancers [53].

In research by Kaur et al., the probable effect of selective agonism by PPAR gamma receptors was studied for radiation therapy in non-small-cell lung carcinoma [54]. The agonist used was Rosiglitazone. A reasonably significant increase in the intensities of radiation-induced apoptosis was detected in H1299 cells attributed to enhanced PPARG expression. Consequently, it was deduced that PPAR gamma agonism stimulates the radio-sensitizing effect.

Another investigation was undertaken on the expression of PPAR gamma in human normal cervix and cervical carcinoma tissues. The effect of PPAR gamma ligands on the sustenance of cervical cancer cells was also an aim. It was observed that the PPAR gamma protein expression, was lessened in cervical carcinoma in comparison to normal cervical tissues [55].

Similar results were revealed using the effect of Ciglitizone on cell proliferation, which reflected noteworthy growth inhibition on human cervical cancer cell lines, C-33-A and C-4II. It further added to the substantial evidence for the role of PPAR in multiple human cervical cancer tissues and cell lines where a downregulation is encountered [56]. Several *in vitro* studies validated those high levels of free fatty acids induce the proliferation, migration, and invasion of prostate cancer cells (PC3 and 22RV1). Therefore, to test the fact, an assessment was done for serum lipid levels in patients suffering from prostate cancer in comparison to normal individuals. It was concluded that high levels of free fatty acids promote cancer by upregulation of expression in PPAR γ [57].

The fact potentiated was that obesity is undoubtedly an important risk factor, resulting in upregulation of PPAR gamma, consequently leading to incidence and progression of PCa [58]. In an interesting work, the expression of PPAR γ was studied in epithelial cells in the colon. There was a differential expression of PPAR in different segments of the colon. Specifically, in the cell lines, Caco-2, and HT-29 human adenocarcinoma cells, PPAR γ expression was amplified upon differentiation. A significant role was observed as reflected in the amplified expression of PPAR γ was observed in the colon (**Table 1**) [73, 74].

S. No.	Bioactive	Mechanism	References
1.	Capsaicin	Induction of apoptosis in melanoma, colon, and prostate cancer cells. It was attributed to the activation of the PPAR γ	[59]
2.	Linoleic acid	It was found to modulate interactions between PPARβ/δ and PPARγ isoforms. Conjugated linoleic acid (CLA) was able to induce apoptosis, upregulate PPARG gene expression and activate PPAR γ protein in certain human cancer cell lines.	[60, 61]
3.	β-carotene	The chemopreventive activity of β-carotene against breast cancer showed a significant increase in PPAR γ mRNA	[62]
4.	Carotenoids: Fucoxanthin	Activates PPAR-γ in cancer cells. It was reported that the edible carotenoid fucoxanthin when combined with 6-troglitazone, induced apoptosis of Caco-2 cells. oreover, in epidemiological studies, the consumption of carotenoids was shown to protect against breast cancer.	[63, 64]
5.	Kaempferol and Apigenin	It was found that flavonoids stimulated PPAR-γ transcriptional activities as allosteric effectors and are beneficial in Prostate cancer.	[65]
6.	Daidzein	They bind to the estrogen-related receptors but also PPARα and PPARγ. As a result, their biological effects are determined by the balance between activated ERs and PPARγ.	[66]
7	Triterpene glycosides: Glycyrrhizin Triterpenoid acid: Betulinic acid	Triterpene glycoside both have shown pro-PPAR-γ activities in cancer cells.	[67]
8	Resveratrol	PPAR γ plays a role in Resveratrol-induced apoptosis of colon carcinoma cells. The combination of resveratrol with a PPARγ agonist rosiglitazone proved as an option for colorectal cancer.	[68]
9	Genistein	It impedes the OS cell cycle as a nontoxic activator of PPAR γ. It has been shown to lower the production of prostaglandin E2 in MDA-MB-231 human breast cancer cells and to reduce the invasiveness of these cells The effect of eicosapentaenoic and docosahexaenoic acids in activating PPAR γ was dependent on genistein.	[69, 70]
10	Flavone Wogonin	It has been shown that PPARα activation by wogonin downregulates osteopontin a multifunctional protein involved in several physiological and pathological events, such as cancer.	[71]
11.	Flavanones Hesperetin, Naringenin, and their glycosides Epigallocatechin-3-galllate	Induced expression of PPAR γ in a dose-dependent manner while naringenin was able to activate PPARα. Increases the expression of PPARα and confers susceptibility to cancer.	[72]

Table 1.
Role of bioactive in the modulation of PPARs for treatment of various cancers.

6. Challenges in anticancer strategies of peroxisomes

To exploit the peroxisomal metabolism for anticancer approaches, several constraints need to be addressed in the first place [75]. Foremost, it should be assessed for its possible side effects on non-malignant cells as overhauling the metabolism might result in serious side effects. Next, the concern might be the differential behavior of the diverse cancer cells and their lineage [76].

The tumor heterogeneity is anyhow allied with differential metabolic activities. In this context, the selection of the study group may be very crucial [77]. It may also affect the prerequisite of peroxisomal functions in a specific subset of tumors [78].

7. Discussion

PPARs, commonly known as modulators of genetic expression, exhibit variant tissue expression depending upon differential microenvironment and have thus attracted a lot of attention whether singly or in a combination strategy [25, 79].

This process can then activate the transcription of various genes involved in diverse physiological and pathophysiological processes that play main roles in the pathogenesis of several chronic diseases, such as atherosclerosis [80], diabetes [81], liver disease [82], cardiovascular diseases [83], and cancer, involving inflammatory effects and their corresponding clinical implications [84].

8. Conclusions

Though therapeutic approaches to target peroxisome metabolism in cancer have been on a rise and pursued very closely by keen researchers, not many modulators have been assessed completely. Henceforth, well-defined *in vivo* models have to be investigated for the potential of peroxisome inactivation to suppress cancer progression. The differential behavior of peroxisomes in different microenvironments will help to facilitate the development of a higher number of effective drugs for the modulation of peroxisomal functions [85].

Peroxisome metabolism has been invariably linked to the functions of organelles, viz. endoplasmic reticulum and mitochondria [86, 87]. The disruption of the association between the organelles and peroxisomes refurbishes the cancer cell metabolism. Further, it can be ascertained, that probable peroxisome targeting with drugs that inhibit the related organelles may lead to amplified anticancer mechanics [75].

It can be useful where targeting peroxisomes might enhance the targeting of other metabolic pathways in cancer. It, however, remains unveiled whether the alteration of peroxisome metabolism is a consequent event due to alterations in metabolism due to cellular changes during cancer or bears a prime position in the development [88].

However, it can never be disregarded that ample research has substantiated the potential of peroxisomes as absolute cancer targets while further exploration role of peroxisome metabolism in the genesis of tumors might prove to be a curtain-raiser. A rational approach to drug design can be attained by the revelation of the regulatory machinery and transcriptional focus of the PPARs. Focused research in this direction may provide a perfect perception of the development of metabolic diseases including cancer.

Acknowledgements

The authors are thankful to the Chancellor, Integral University, Lucknow-226026, Uttar Pradesh, India for his sustained encouragement, meticulous supervision, and valuable suggestions at all stages of completion of this chapter. The authors are thankful to the Faculty of Pharmacy, Integral University for providing all the necessary facilities related to the present work. The authors are also thankful for the research cell, Integral University.

Conflict of interest

The authors declare no conflict of interest.

Author details

Shazia Usmani[1*], Shadma Wahab[2], Abdul Hafeez[1], Shabana Khatoon[1] and Syed Misbahul Hasan[1]

1 Faculty of Pharmacy, Integral University, Lucknow, India

2 Department of Pharmacognosy, College of Pharmacy, King Khalid University, Abha, Saudi Arabia

*Address all correspondence to: shazia@iul.ac.in

References

[1] Delille HK, Bonekamp NA, Schrader M. Peroxisomes and disease: An overview. International Journal of Biomedical Science. 2006;**2**(4):308-314

[2] Daich Varela M, Jani P, Zein WM. The peroxisomal disorder spectrum and Heimler syndrome: Deep phenotyping and review of the literature. American Journal of Medical Genetics C Seminar Medical Genetics. 2020;**184**(3):618-630

[3] Joshi AS, Cohen S. Lipid droplet and peroxisome biogenesis: Do they go hand-in-hand. Frontiers in Cell Development Biology. 2019;**7**:92

[4] Van der Klei IJ. Peroxisomes. In: McManus LM, Mitchell RN, editors. Pathobiology of Human Disease: A Dynamic Encyclopedia of Disease Mechanisms. Elsevier; 2014. pp. 108-113

[5] Lismont C, Revenco I, Fransen M. Peroxisomal hydrogen peroxide metabolism and signaling in health and disease. International Journal of Molecular Sciences. 2019;**20**(15):3673

[6] Cipolla CM, Lodhi IJ. Peroxisomal dysfunction in age-related diseases. Trends in Endocrinology and Metabolism. 2017;**28**(4):297-308

[7] Schrader M, Grille S, Fahimi HD, Islinger M. Peroxisome interactions and cross-talk with other subcellular compartments in animal cells. Sub-cellular Biochemistry. 2013;**69**:1-22

[8] Lodhi IJ, Semenkovich CF. Peroxisomes: A nexus for lipid metabolism and cellular signaling. Cell Metabolism. 2014;**19**(3):380-392

[9] Bhatti JS, Bhatti GK, Reddy PH. Mitochondrial dysfunction and oxidative stress in metabolic disorders: A step towards mitochondria based therapeutic strategies. Biochimica et biophysica acta. 1863;**5**:1066-1077

[10] Ferdinandusse S, Denis S, Faust PL, Wanders RJ. Bile acids: The role of peroxisomes. Journal of Lipid Research. 2009;**50**(11):2139-2147

[11] Uzor NE, McCullough LD, Tsvetkov AS. Peroxisomal dysfunction in neurological diseases and brain aging. Frontiers in Cell Neuroscience. 2020;**14**:44

[12] Gonzalez H, Hagerling C, Werb Z. Roles of the immune system in cancer: From tumor initiation to metastatic progression. Genes & Development. 2018;**32**(19-20):1267-1284

[13] Dahabieh MS, Di Pietro E, Jangal M, Goncalves C, Witcher M, Braverman NE, et al. Peroxisomes and cancer: The role of a metabolic specialist in a disease of aberrant metabolism. Biophysica. 2018;**1870**(1):103-121

[14] Tyagi S, Gupta P, Kaushal SAS, C, Sharma S. The peroxisome proliferator-activated receptor: A family of nuclear receptors role in various diseases. Journal of Advanced Pharmaceutical Technology & Research. 2011;**2**(4):236-240

[15] Kersten S. Integrated physiology and systems biology of PPARα. Molecular Metabolism. 2014;**3**(4):354-371

[16] Peters JM, Shah YM, Gonzalez FJ. The role of peroxisome proliferator-activated receptors in carcinogenesis and chemoprevention. Nature Reviews: Cancer. 2012;**12**(3):181-195

[17] Berger J, Moller DE. The mechanisms of action of PPARs. Annual Review Medicine. 2002;**53**:409-435

[18] Ahmadian M, Suh JM, Hah N. PPARγ signaling and metabolism: The good, the bad and the future. Natural Medicine. 2013;**19**(5):557-566

[19] Isseman I, Green S. Activation of a member of the steroid hormone receptor superfamily by peroxisome proliferators. Nature. 1990;347(6294):645-654

[20] Grygiel-Gorniak B. Peroxisome proliferator-activated receptors and their ligands: Nutritional and clinical implications-a review. Nutrition Journal. 2014;**13**:17

[21] Feige JN, Gelman L, Michalik L, Desvergne B, Wahli W. From molecular action to physiological outputs peroxisome proliferator-activated receptors are nuclear receptors at the crossroads of key cellular functions. Progress in Lipid Research. 2006;**45**(2):120-159

[22] Shi Y, Hon M, Evans RM. The peroxisome proliferator activated receptor, an integrator of transcriptional repression and nuclear receptor signaling. Proceedings of the National Academy of Sciences of the United States of America. 2002;**99**(5):2613-2618

[23] Greene ME, Blumberg B, McBride OW, Yi HF, Kronquist K, Kwan K, et al. Isolation of the human peroxisome proliferator activated receptor gamma cDNA: Expression in hematopoietic cells and chromosomal mapping. Gene Expression. 1995;**4**(4-5): 281-299

[24] Elbrecht A, Chen Y, Cullinan CA, Hayes N, Leibowitz MD, Moller DE, et al. Molecular cloning, expression and characterization of human peroxisome proliferator activated receptors gamma 1 and gamma 2. Biochemical and Biophysical Research Communications. 1996;**224**(2):431-437

[25] Michalik L, Auwerx J, Berger JP, Chatterjee VK, Glass CK, Gonzalez FJ, et al. International union of pharmacology LXI. Peroxisome proliferator-activated receptors. Pharmacological Reviews. 2006;**58**(4):726-741

[26] Suwaki N, Masuyama H, Masumoto A, Takamoto N, Hiramatsu Y. Expression and potential role of peroxisome proliferator-activated receptor gamma in the placenta of diabetic pregnancy. Placenta. 2007;**28**(4):315-323

[27] Ezzeddini R, Taghikhani M, Salek Farrokhi A, Somi MH, Samadi N, Esfahani A, et al. Downregulation of fatty acid oxidation by involvement of HIF-1and PPAR in human gastric adenocarcinoma and its related clinical significance. Journal of Physiology and Biochemistry. 2021;77(2):249-260

[28] Lehrke M, Lazar MA. The many faces of PPARgamma. Review Cell. 2005;**123**(6):993-999

[29] Khateeb J, Gantman A, Kreitenberg AJ, Aviram M. Fuhrman B Paraoxonase 1 (PON1) expression in hepatocytes is upregulated by pomegranate polyphenols: A role for PPAR -gamma pathway. Atherosclerosis. 2010;**208**(1):119-125

[30] Rochlani Y et al. Metabolic syndrome: Pathophysiology, management, and modulation by natural compounds. Therapeutic Advances in Cardiovascular Disease. 2017;**11**(8):215-225

[31] Azhar S. Peroxisome proliferator-activated receptors, metabolic syndrome and cardiovascular disease. Future Cardiology. 2010;**6**(5):657-691

[32] Mueller M, Beck V, Jungbauer A. PPAR α activation by culinary herbs and spices. Planta medica. 2010;77:497-504

[33] Salehi B, Machin L, Monzote L, Ezzat SM, Salem MA, Merghany M, et al. Therapeutic potential of quercetin: New insights and perspectives for human health. ACS Omega. 2020;**5**(20):11849-11872

[34] Ahmed M, Daisuke Y, Masahiko K, Susumu K. Therapeutic strategy for targeting aggressive malignant gliomas by disrupting their energy balance. Journal of Biological Chemistry. 2016;**291**(41):21496-21509

[35] Grabacka M, Reiss K. Anticancer properties of PPAR alpha-effects on cellular metabolism and inflammation. PPAR Research. 2008;**2008**:930705

[36] Muzio G, Barrera G, Pizzimenti S. Peroxisome proliferator-activated receptors (PPARs) and oxidative stress in physiological conditions and in cancer. Antioxidants. 2021;**10**(11):1734

[37] Qie S, Diehl JA, Cyclin D1. Cancer progression, and opportunities in cancer treatment. Journal of Molecular Medicine. 2016;94(12):1313-1326

[38] Augimeri G, Giordano C, Gelsomino L, et al. The role of PPARγ ligands in breast cancer: From basic research to clinical studies. Cancers (Basel). 2020;**12**(9):2623

[39] Dixit G. AratiPrabhu, The pleiotropic peroxisome proliferator activated receptors. Regulation and therapeutics. Experimental and Molecular Pathology. 2022;**124**:104723

[40] Koeffler HP. Peroxisome proliferator-activated receptor gamma and cancers. Clinical Cancer Research. 2003;**9**(1): 1-9

[41] Morinishi T, Tokuhara Y, Ohsaki H, Ibuki E, Kadota K, Hirakawa E. Activation and expression of peroxisome proliferator-activated receptor alpha are associated with tumorigenesis in colorectal carcinoma. Peroxisome Proliferator-activated Research. 2019;**3**:7486727

[42] Park JI, Kwak JY. The role of peroxisome proliferator-activated receptors in colorectal cancer. Peroxisome Proliferator-activated Research. 2012;**2012**:876418

[43] Jiang J, Xie Z, Guo J, et al. Association of *PPARG* polymorphism with risk of colorectal cancer: From a case-control study to a meta analysis. Oncotarget. 2017;**8**(59):100558-100569

[44] Wagner N, Wagner KD. PPAR Beta/Delta and the hallmarks of cancer. Cells. 2020;**9**(5):1133

[45] Ikezoe T, Miller CW, Kawano S, et al. Mutational analysis of the peroxisome proliferator-activated receptor γ in human malignancies. Cancer Research. 2001;**61**(13): 5307-5310

[46] Kubota T, Koshizuka K, Williamson EA, et al. Ligand for peroxisome proliferator-activated receptor γ (troglitazone) has potent antitumor effect against human prostate cancer both in vitro and in vivo. Cancer Research. 1998;**58**(15):3344-3352

[47] De la Lastra CA, Sanchez-Fidalgo S, Villegas I, Motilva V. New pharmacological perspectives and therapeutic potential of PPAR-γ agonists. Current Pharmaceutical Design. 2004;**10**(28):3505-3524

[48] Keller JM, Cable S, Armbruster L, Ciolek E, Colin S, Schilt J, et al. Peroxisome through cell differentiation and neoplasia. Biology Cell. 1993;*77*:77-88

[49] Lauer C, Volkl A, Riedl S, Fahimi HD, Beier K. Impairment of peroxisomal biogenesis in human colon carcinoma. Carcinogenesis. 1999;**20**:985-989

[50] Litwin JA, Beier K, Völkl A, Hofmann WJ, Fahimi HD. Immunocytochemical investigation of catalase and peroxisomal lipid beta-oxidation enzymes in human hepatocellular tumors and liver cirrhosis. Virchows Arch. 1999;**435**:486-495

[51] Frederiks WM, Bosch KS, Hoeben KA, Van Man J, Langbein S. Renal cell carcinoma and oxidative stress: The lack of peroxisomes. Acta Histochem. 2010;**112**:364-371

[52] Kim JA. Peroxisome metabolism in cancer. Cells. 2020;**9**(7):1692

[53] Fransen M, Nordgren M, Wang B, Apanasets O. Role of peroxisomes in ROS/RNS-metabolism: Implications for human disease. Biochimica Biophysica Acta. 2012;**1822**:1363-1373

[54] Heudobler D, Rechenmacher M, Lüke F, et al. Peroxisome proliferator-activated receptors (PPAR)γ agonists as master modulators of tumor tissue. International Journal of Molecular Sciences. 2018;**19**(11):3540

[55] Kaur S, Nag A, Gangenahalli G, Sharma K. Peroxisome proliferator activated receptor gamma sensitizes non-small cell lung carcinoma to gamma irradiation induced apoptosis. Frontiers in Genetics. 2019;**13**(10):554

[56] Zhang S, Liu F, Mao X, Huang J, Yang J, Yin X, et al. Elevation of miR-27b by HPV16 E7 inhibits PPARγ expression and promotes proliferation and invasion in cervical carcinoma cells. International Journal of Oncology. 2015;**47**(5):1759-1766

[57] Plissonnier ML, Fauconnet S, Bittard H, Mougin C, Rommelaere J, Lascombe I. Cell death and restoration of TRAIL-sensitivity by ciglitazone in resistant cervical cancer cells. Oncotarget. 2017;**8**(64):107744-107762

[58] Ha X, Wang J, Chen K, et al. Free fatty acids promote the development of prostate cancer by upregulating peroxisome proliferator-activated receptor gamma. Cancer Management and Research. 2020;**12**:1355-1369

[59] Yang KM, Pyo JO, Kim GY, Yu R, Han IS, Ju SA, et al. Capsaicin induces apoptosis by generating reactive oxygen species and disrupting mitochondrial transmembrane potential in human colon cancer cell lines. Cellular Molecular Biology Letters. 2009;**14**(3):497-510

[60] Słowikowski BK, Drzewiecka H, Malesza M, Mądry I, Sterzyńska K, Jagodziński PP. The influence of conjugated linoleic acid on the expression of peroxisome proliferator-activated receptor-γ and selected apoptotic genes in non-small cell lung cancer. Molecular and Cellular Biochemistry. 2020;**466**(1-2):65-82

[61] Sasaki T, Yoshida K, Shimura H, Ichiba M, Sasahira T, Shimomoto T, et al. Inhibitory effect of linoleic acid on transformation of IEC6 intestinal cells by in vitro azoxymethane treatment. International Journal of Cancer. 2006;**118**(3):593-599

[62] Cui Y, Lu Z, Bai L, Shi Z, Zhao WE, Zhao B. beta-Carotene induces apoptosis and up-regulates peroxisome proliferator-activated receptor gamma expression and reactive oxygen species production in MCF-7 cancer cells. European Journal of Cancer. 1990;**43**(17):2590-2601

[63] Hosokawa M, Kudo M, Maeda H, Kohno H, Tanaka T, Miyashita K.

Fucoxanthin induces apoptosis and enhances the antiproliferative effect of the PPARgamma ligand, troglitazone, on colon cancer cells. Biochimica Biophysica Acta. 2004;**18**:113-119

[64] Dorgan JF, Sowell A, Swanson CA, et al. Relationships of serum carotenoids, retinol, α-tocopherol, and selenium with breast cancer risk: Results from a prospective study in Columbia, Missouri (United States). Cancer Causes & Control. 1998;**9**(1):89-97

[65] Sainis I, Vareli K, Karavasilis V, Briasoulis E. PPARgamma: The portrait of a target ally to cancer chemopreventive agents. Peroxisome Proliferator-activated Research. 2008;**2008**:436489

[66] Lapillonne H, Konopleva M, Tsao T, Gold D, Mcqueen T, Madden T, et al. Activation of peroxisome proliferator-activated receptor-gamma by a novel synthetic triterpenoid CDDO induces growth arrest and apoptosis in breast cancer cells. International Journal of Cancer. 2003;**66**

[67] Aires V, Brassart B, Carlier A, Scagliarini A, Mandard S, Limagne E, et al. A role for peroxisome proliferator-activated receptor gamma in resveratrol-induced colon cancer cell apoptosis. Molecular Nutrition Food Research. 2014;**58**(9):1785-1794

[68] Song M, Tian X, Lu M, Zhang X, Ma K, Lv Z, et al. Genistein exerts growth inhibition on human osteosarcoma MG-63 cells via PPARγ pathway. International Journal Oncology. 2015;**46**(3):1131-1140

[69] Patel RP, Barnes S. Isoflavones and PPAR signaling: A critical target in cardiovascular, metastatic, and metabolic disease. Peroxisome Proliferator-activated Research. 2010;**2010**: 153252

[70] Zhang YM, Li MX, Tang Z, Wang CH. Wogonin suppresses osteopontin expression in adipocytes by activating PPARα. Acta Pharmacology. 2015;**36**:987-997

[71] Daniela R, Carmina S, Orazio T. The potential of natural products for targeting PPARα. Acta Pharmaceutica Sinica. 2017;**7**:4

[72] Oltedal S, Skaland I, Maple J, et al. Expression profiling and intracellular localization studies of the novel Proline-, Histidine-, and Glycine-rich protein 1 suggest an essential role in gastro-intestinal epithelium and a potential clinical application in colorectal cancer diagnostics. Biomedicine Medical Central. 2018;**18**:26

[73] Ayala A, Muñoz MF, Argüelles S. Lipid peroxidation: Production, metabolism, and signaling mechanisms of malondialdehyde and 4-hydroxy-2-nonenal. Oxidative Medicine Cell Longevity. 2014;**2014**:360438

[74] Mastropietro G, Tiscornia I, Perelmuter K, Astrada S, Bollati-Fogolín M. HT-29 and Caco-2 reporter cell lines for functional studies of nuclear factor Kappa B activation. Mediators of Inflammation. 2015;**13**:860534

[75] Luengo A, Gui DY, Vander Heiden MG. Targeting metabolism for cancer therapy. Cell Chemical Biology. 2017;**24**(9):1161-1180

[76] Gentric G, Mieulet V, Mechta-Grigoriou F. Heterogeneity in cancer metabolism: New concepts in an old field. Antioxidants & Redox Signaling. 2017;**26**:462-485

[77] Tasdogan A, Faubert B, Ramesh V, Ubellacker JM, Shen B, et al. Metabolic heterogeneity confers differences in

melanoma metastatic potential. Nature. 2020;**577**:115-120

[78] Kim S, Lee Y, Koo JS. Differential expression of lipid metabolism-related proteins in different breast cancer subtypes. Public Library of Science one. 2015;**10**:82

[79] Kersten S, Mandard S, Escher P, Gonzalez FJ, Tafuri S, Desvergne B, et al. The peroxisome proliferator-activated receptor alpha regulates amino acid metabolism. Federation of American Societies for Experimental Biology. 2001;**15**(11):1971-1976

[80] Marx N, Bourcier T, Sukhova GK, Libby P, Plutzky J. PPARgamma activation in human endothelial cells increases plasminogen activator inhibitor type-1 expression: PPARgamma as a potential mediator in vascular disease. Arteriosclerosis, Thrombosis and Vascular Biology. 1999;**19**(3):546-551

[81] Neschen S, Morino K, Dong J, Wang-Fischer Y, Cline GW, Romanelli AJ, et al. Fatty acids preserve insulin sensitivity in a peroxisome proliferator-activated receptor-alpha-dependent manner. Diabetes. 2007;**56**(4):1034-1041

[82] Rao MS, Reddy JK. PPAR alpha in the pathogenesis of fatty liver disease. Hepatology. 2004;**40**(4):783-786

[83] Takano H, Komuro I. Peroxisome proliferator-activated receptor gamma and cardiovascular diseases. Circulation Journal. 2009;**73**(2):214-220

[84] Moraes LA, Piqueras L, Bishop-Bailey D. Peroxisome proliferator-activated receptors and inflammation. Pharmacology and Therapeutics. 2006;**110**(3):371-385

[85] Ganguli G, Mukherjee U, Sonawane A. Peroxisomes and oxidative stress: Their implications in the modulation of cellular immunity during mycobacterial infection. Frontier Microbiology. 2019;**10**:1121

[86] Wanders RJ, Waterham HR, Ferdinandusse S. Metabolic interplay between peroxisomes and other subcellular organelles including mitochondria and the endoplasmic reticulum. Frontier Cell Division Biology. 2015;**3**:83

[87] Fransen M, Lismont C, Walton P. The peroxisome-mitochondria connection: How and why? International Journal Molecular Science. 2017;**18**:1126

[88] Xia M, Zhang Y, Jin K. Communication between mitochondria and other organelles: A brand-new perspective on mitochondria in cancer. Cell Bioscience. 2019;**9**:27

Chapter 6

Effect of GW9662 and T0070907 Antagonist of PPARg and Their Coadministration Pairwise with Obestatin on Lipid Profile of DIO-C57BL/6 Mice

Beekanahalli G. Mallikarjuna and Uma V. Manjappara

Abstract

Obestatin and its fragment analog Nt8U were shown to upregulate glycerolipid metabolism and PPARg signaling and decrease fat accumulation in Swiss albino mice. It was further investigated if these peptides could decrease lipid accumulation under obese conditions. We chose to work on Diet-Induced-Obese (DIO) C57BL/6 mice to study the same. Both obestatin and Nt8U decreased lipid accumulation in DIO-C57BL/6 mice. PPARg was not upregulated in comparison to 60% high-fat diet (HFD) fed control mice, implying there was already enhanced PPARg expression due to HFD consumption. We also wanted to investigate if upregulation of PPARg signaling was a secondary effect of enhanced glycerolipid metabolism. To investigate the same, we administered obestatin pairwise with 2 agonists and 2 antagonists of PPARg. The results revealed obestatin is not a mere agonist of PPARg but can also decrease lipid accumulation brought about by rosiglitazone, a well-studied agonist of PPARg. The antagonists also show a further decrease in lipid accumulation, probably due to inhibition of PPARg activity brought about by HFD and the additive decrease brought about by obestatin in DIO-C57BL/6 mice. This chapter will be structured to briefly introduce obestatin, Nt8U, their effect on gene expression in the adipose tissue, and the effect of PPARg agonists and antagonists on their ability to reduce fat accumulation.

Keywords: DIO-C57BL/6, obestatin, rosiglitazone, GW9662, T0070907

1. Introduction

Obestatin is a regulatory peptide discovered in 2005 by Zhang et al. It was predicted to be a protease cleavage product of preproghrelin. The 23-residue peptide was isolated and characterized for its activity. It was shown to decrease food intake and body weight in rodents. They also continued to show GPR39 an orphan G-protein coupled receptor could be the cognate receptor for obestatin [1]. These claims have been debated upon since [2]. More recent research shows GLP-1 receptor could be mediating its activity [3]. In our laboratory, we synthesized 3 overlapping fragments of obestatin and showed the N-terminal 13 residues mimicked the parent peptide obestatin the best [4]. Subsequently, we synthesized two analogs of the N-terminal peptide with alpha-aminoisobutyric acid (Aib, denoted as U) at position 8 replacing a glycine (Nt8U) and cyclohexyl amino acid (Cha) at position 5 replacing phenylalanine (Nt5Cha). Experiments in Swiss albino mice showed Nt8U to be as active as obestatin in its ability to reduce fat accumulation [5]. Obestatin and its fragment analog Nt8U were shown to upregulate glycerolipid metabolism and PPARg signaling and decrease fat accumulation in Swiss albino mice [6]. It was further investigated if these peptides could decrease lipid accumulation under obese conditions. We chose to work on Diet-Induced-Obese (DIO) C57Bl/6 mice to study the same. Both Obestatin and Nt8U decreased lipid accumulation in DIO-C57BL/6 mice. PPARg was not upregulated in comparison to 60% high-fat diet (HFD) fed control mice, implying there was already enhanced PPARg expression due to HFD consumption [7, 8].

We also wanted to investigate if upregulation of PPARg signaling was a secondary effect of enhanced glycerolipid metabolism. To investigate the same, we administered obestatin pairwise with 2 agonists and 2 antagonists of PPARg. The results revealed obestatin is not a mere agonist of PPARg but can also decrease lipid accumulation brought about by rosiglitazone, a well-studied agonist of PPARg [9]. The antagonists also show a further decrease in lipid accumulation, probably due to inhibition of PPARg activity brought about by HFD and the additive decrease brought about by obestatin in DIO-C57BL/6 mice [5]. The effect of antagonists of PPARg in DIO-C57BL/6 mice will be discussed in this chapter concerning DIO and Rosiglitazone (PPARg agonist) + obestatin administered DIO-C57BL/6 mice.

2. Effect of antagonists of PPARg, GW9662, and T0070907 on DIO-C57BL/6 mice individually and along with obestatin

GW9662 is a potent and selective PPARγ antagonist with an IC_{50} of 3.3 nM. It has 10 and 600 fold less selectivity towards PPARα and PPARδ respectively. Mass spectrometric analysis revealed Cys285 was covalently modified by GW9662 [10]. T0070907 was identified as a potent and selective PPAR antagonist. It had an apparent binding affinity of 1 nM. It covalently modifies cysteine 313 in helix 3 of human PPAR2. **Figure 1** shows the structure of the antagonists [11].

GW9662 T0070907

Figure 1.
Structure of the PPARg antagonists GW9662 and T0070907.

2.1 Induction of obesity and mice experiments

Obesity was induced by administering a 60% calorie by high-fat diet (HFD) for a period of 24 weeks to four-week-old male C57BL/6 mice. The gain in weight over the induction period is shown in **Figure 2a** as a comparison with normal diet-fed male C57BL/6 mice. The HFD fed mice gained an average of 25 g whereas, the normal chow-fed mice gained 5 g of weight over the same period. Subsequently, the mice were grouped into six groups as follows:

Group 1: HFD Control.

Group 2: Obestatin Control.

Group 3: GW9662 Control.

Group 4: T0070907 Control.

Group 5: Obestatin + GW9662 treatment.

Group 6: Obestatin + T0070907 treatment.

Obestatin was synthesized, purified, and characterized in our laboratory as described previously [4]. GW9662 and T0070907 were purchased from Sigma Aldrich. As per the previous optimization done in our laboratory obestatin was administered at 160 nmol /kg/BW [12]. From the available literature, it was decided GW9662 and T0070907 should be administered at 1 mg/kg/BW to the respective groups [10, 11]. All samples were dissolved in 20% DMSO in 0.9% saline and the same solvent was used as control. All the administrations were intraperitoneal. The mice were administered the respective compound for 8 days after induction of obesity. Food intake was monitored for 5 h on an hourly basis after administration of saline or the respective compounds and gain in body weight was recorded every day. **Figure 3** shows the decrease in food intake upon administration of obestatin followed by further decreased food intake by the T0070907 group, Obestatin + T0070907 group, GW9662 group, and Obestatin + GW9662 group, respectively. **Figure 2b** shows the gain in body weight during 8 days. It can be seen that all the treated groups showed a negative gain in body weight even upon HFD administration. Only the HFD control group showed a steady gain in body weight. After 8 days of experimentation, the mice fasted for 6 hours and blood were

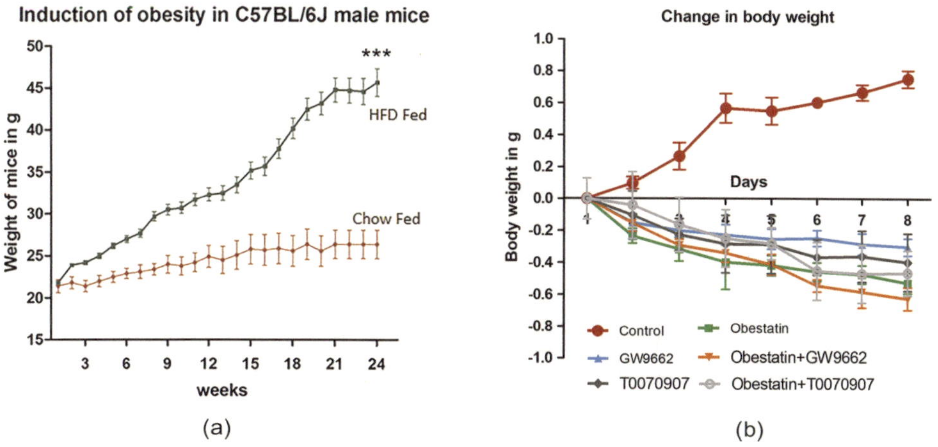

Figure 2.
(a) Obesity induction over a period of 24 weeks in comparison to Normal diet fed mice. (b) Change in body weight upon treatment with the respective compounds.

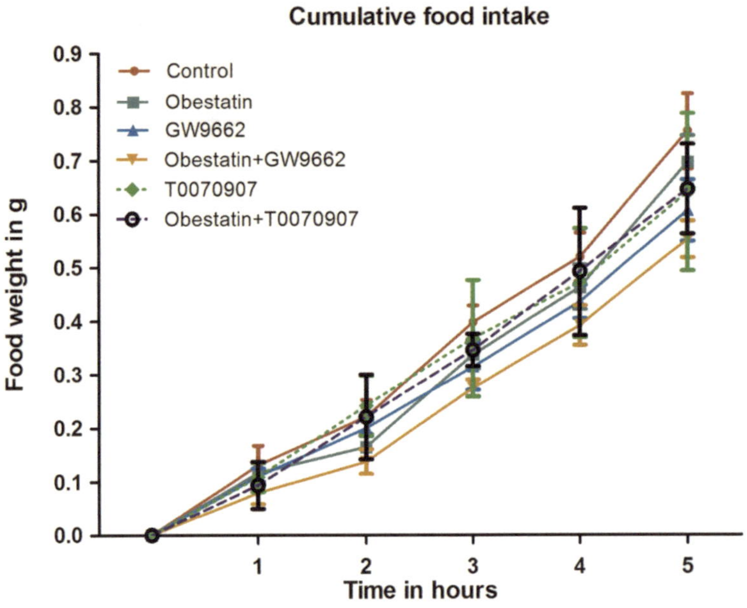

Figure 3.
Food intake for 5 h after administration of the respective compounds. The mice were administered the respective compounds and fed HFD after 15 min.

drawn through the retro-orbital plexus and the mice were sacrificed according to established protocols. All adipose depots and vital organs were stored in formalin for histopathology studies and part of them were frozen in liquid nitrogen for RNA extraction and profiling.

2.2 Effect of the treatments on the different fat pad

Epididymal, perirenal, retroperitoneal, inguinal, BAT, gluteal, axillary, and cervical fat pads were collected from each group and weighed and normalized to the respective bodyweight to access the effect of the treatments on fat accumulation in each adipose depot. As not all adipose depots are metabolically the same, a decrease in fat accumulation in certain depots can be more beneficial. **Table 1** summarizes the overall effect of the treatment on the fat depots. GW9662 increased epididymal adipose tissue weight in comparison to the HFD and obestatin treated groups and decreased % epididymal fat upon coadministration with obestatin. It significantly decreased % Inguinal fat weight upon coadministration with obestatin by 26.4%. It also decreased total subcutaneous fat by 11.3% but increased % visceral fat by 15% and thereby increased total fat % in the adipose depots by 11.1%. On the other hand, T0070907 increased % epididymal fat by 24% and % visceral fat by 10%. It decreased perirenal fat by 54% and % subcutaneous fat by 10%. Upon coadministration with obestatin, T0070907 decreased % inguinal fat by 27% and % total subcutaneous fat by 15%. And maintaining total fat content is equal to that of the HFD control.

2.3 Plasma biochemical analysis and lipid parameters

SGOT, SGPT, alkaline phosphatase (ALP), fasting blood glucose, creatinine, urea, triglyceride, total cholesterol (TC), and high-density lipoprotein cholesterol

Plasma lipid parameters (mg/dl)	Control	Obestatin	GW9662	Obestatin + GW9662	T0070907	Obestatin + T0070907
% Epididymal fat wt.	4.771 ± 0.08	4.445 ± 0.12	7.029 ± 0.19$^{*\#}$	6.193 ± 0.12$^{*\#}$	5.914 ± 0.18$^{*\#}$	5.474 ± 0.40#
% Perirenal fat wt.	0.330 ± 0.02	0.378 ± 0.03	0.142 ± 0.02$^{*\#}$	0.180 ± 0.03$^{*\#}$	0.149 ± 0.03$^{*\#}$	0.182 ± 0.02*#
% Retroperitoneal fat wt.	2.608 ± 0.05	2.564 ± 0.07	2.414 ± 0.15	2.446 ± 0.13	2.356 ± 0.20	2.244 ± 0.14
% Inguinal fat wt.	5.119 ± 0.11	4.877 ± 0.08	4.285 ± 0.70	3.768 ± 0.70$^{*\#}$	4.358 ± 0.67	3.739 ± 0.54*
% BAT wt.	0.506 ± 0.01	0.545 ± 0.03	0.430 ± 0.04$^{\#}$	0.428 ± 0.03$^{\#}$	0.354 ± 0.02$^{*\#}$	0.420 ± 0.02#
% Gluteal Fat wt.	1.250 ± 0.06	1.530 ± 0.03	1.326 ± 0.06	1.586 ± 0.10*	1.269 ± 0.09	1.335 ± 0.10
% Axillary fat wt.	1.012 ± 0.02	1.018 ± 0.04	1.145 ± 0.06	1.207 ± 0.05	1.149 ± 0.05	1.163 ± 0.04
% Cervical fat wt.	0.447 ± 0.01	0.376 ± 0.01	0.447 ± 0.03	0.406 ± 0.04	0.404 ± 0.04	0.386 ± 0.02
% Total visceral fat wt.	7.6480 ± 0.05	7.388 ± 0.07	9.585 ± 0.09$^{*\#}$	8.819 ± 0.23$^{*\#\$}$	8.419 ± 0.10$^{*\#}$	7.900 ± 0.19
% Total subcutaneous fat wt.	8.335 ± 0.05	8.346 ± 0.04	7.633 ± 0.11$^{*\#}$	7.396 ± 0.09$^{*\#}$	7.535 ± 0.09$^{*\#}$	7.044 ± 0.07$^{*\#@}$
% Total fat wt.	15.497 ± 0.08	14.824 ± 0.07*	15.63 ± 0.14$^{\#}$	17.218 ± 0.09$^{*\#\$}$	15.954 ± 0.09$^{*\#}$	14.944 ± 0.13$^{*@}$

*Significant when compared to control.
#Significant when compared to obestatin.
$Significant when compared to GW9662.
@Significant when compared toT0070907).

Table 1.
Effect of obestatin (160 nmol /kg/BW), GW9662 (1 mg/kg/BW), T0070907 (1 mg/kg/BW), individually and in combination of obestatin + GW9662 (160 nmol/kg/BW + 1 mg/kg/BW, obestatin + T0070907 (160 nmol/kg/BW + 1 mg/kg/BW) treatment on % fat pad weight. Data are expressed as the mean ± SEM (N≧8). Data are expressed as the mean ± SEM (N≧8). P < 0.05 was considered as statistically significant value.

(HDL-C) were estimated using commercially available kits. Phospholipid estimation was carried out by a colorimetric method [13]. Plasma Leptin and adiponectin concentrations were tested as per the instructions of the manufacturer of commercially available ELISA kits.

Table 2 summarizes the plasma lipid parameters after the 8 days of the experiment. All the treated groups showed a significant decrease in plasma triglyceride. Obestatin + GW9662 and obestatin + T0070907 showed a maximum decrease of 39%. T0070907 and obestatin + T0070907 showed a maximum decrease in plasma total cholesterol by 10% whereas, GW9662 showed a significant decrease in total cholesterol by 4% only upon coadministration with obestatin. Obestatin increased phospholipids by 32%, the coadministered groups show a weak enhancement indicating there is a combined effect and not that of the individual components. Plasma-free fatty acids are decreased in all treated groups. A maximum decrease of 23% is seen in the obestatin + GW9662 treated group, followed by obestatin + T0070907 by 17%, GW9662 by 14%, T0070907, and obestatin by 10%.

Table 3 summarizes the plasma biochemical parameters. Plasma glucose, protein, urea, and creatinine are in the normal range, did not show any significant changes. Marker enzymes SGOT, SGPT, and ALP are in the normal range for all the groups. Adipokine leptin that signals long-term fat reserves to the brain is significantly decreased most by obestatin + GW9662 group by 45% followed by obestatin + T0070907 by 37%, GW9662 by 33%, and T0070907 by 21% indicating a decrease in fat content. There were no significant changes in adiponectin levels.

2.4 Adipose tissue lipid parameters

Liver tissue, inguinal, and epididymal fat lipids were extracted by Folch's method of lipid extraction [14]. Tissue TAG, TC, and phospholipid were estimated by a colorimetric method [13, 15, 16]. **Table 4** summarizes the tissue lipid parameters. In epididymal adipose tissue, obestatin + T0070907 decreased triglyceride by 20%, followed by obestatin + GW9662 by 17% and GW9662 by 10%. Phospholipids were increased by about 80% in all the groups. No significant decrease in total cholesterol was observed. In inguinal adipose tissue, a significant decrease by 23% in total cholesterol was observed in both obestatin + GW9662 and obestatin + T0070907 groups followed by T0070907 at 18%. No significant changes were observed in triglyceride or phospholipids levels. In the liver tissue, a significant increase of 44% was seen only with respect to triglyceride in the obestatin treated group compared to that of the HFD control.

2.5 mRNA profiling of epididymal and inguinal adipose tissue of lipid metabolism-related genes by quantitative real time-PCR (qPCR)

Total RNA was isolated from the adipocyte cells using TRIzol reagent from Sigma, USA. The quantity and quality of the isolated RNA were assessed using a microspectrophotometer (Eppendorf). Samples having a ratio of A260/280 > 1.8 were used for cDNA synthesis by kit method (Thermo Scientific, Ltd.). Real-time PCR assays were performed using SYBR Green (BioRad CFX96 Touch Real-Time PCR Detection System) and primer sequences of the respective genes are given in SI **Table 1**. Fold change in gene expression was tabulated by normalizing the values of threshold cycle (CT) of the target gene with the CT value of housekeeping gene GAPDH. Briefly, the fold changes were calculated using the $2-\Delta\Delta Ct$ calculation

Plasma lipid parameters (mg/dl)	Control	Obestatin	GW9662	Obestatin + GW9662	T0070907	Obestatin + T0070907
Triglycerides	100.40 ± 2.97	79.90 ± 1.59*	82.29 ± 2.87*#	60.70 ± 3.12*$	73.67 ± 1.41*	59.24 ± 1.54*#@
Total cholesterol	192.40 ± 1.73	196.90 ± 2.35	204.90 ± 3.18*	185.10 ± 2.84#$	173.40 ± 2.75*#	170.40 ± 2.18*#
HDL-cholesterol	146.90 ± 1.66	163.60 ± 1.86*	172.61 ± 2.70*	177.54 ± 3.06*#	121.54 ± 2.62*#	139.54 ± 2.95*#@
Phospholipids	34.62 ± 0.79	45.69 ± 1.09*	34.56 ± 1.41#	37.68 ± 1.53#	33.87 ± 1.09#	36.90 ± 1.84#
Free fatty acids	146.90 ± 1.56	132.20 ± 1.15*	126.45 ± 2.28*	112.56 ± 2.66*#	132.50 ± 2.39*	122.20 ± 1.14*#@

*Significant when compared to control.
#Significant when compared to obestatin.
$Significant when compared to GW9662.
@Significant when compared toT0070907.

Table 2.
Effect of obestatin (160 nmol /kg/BW), GW9662 (1 mg/kg/BW), T0070907 (1 mg/kg/BW), individually and in combination of obestatin + GW9662 (160 nmol /kg/BW + 1 mg/kg/BW, obestatin + T0070907 (160 nmol /kg/BW + 1 mg/kg/BW) treatment on fasting plasma lipid parameters such as triglycerides, total cholesterol, HDL-cholesterol, and phospholipids level. Data are expressed as the mean ± SEM (N≧8). P < 0.05 was considered as statistically significant value.

Fasting plasma parameters	Control	Obestatin	GW9662	Obestatin + GW9662	T0070907	Obestatin + T0070907
Glucose (mg/dl)	166.33 ± 5.15	171.33 ± 5.30	179.80 ± 3.45	159.00 ± 3.06$^{\$}$	176.8 ± 2.74	168.30 ± 3.10
Protein (g/dl)	4.78 ± 0.14	4.50 ± 0.12	4.76 ± 0.19	4.67 ± 0.16	4.47 ± 0.12	4.54 ± 0.19
Urea (mg/dl)	59.33 ± 6.07	55.98 ± 6.27	56.87 ± 3.41	50.68 ± 3.07	56.87 ± 2.25	53.87 ± 1.41
Creatinine (g/dl)	2.48 ± 0.18	2.57 ± 0.19	2.76 ± 0.16	2.65 ± 0.07	2.78 ± 0.09	2.76 ± 0.08
SGOT (U/L)	19.80 ± 1.29	18.06 ± 1.16	16.78 ± 1.32	19.76 ± 1.36	17.54 ± 1.01	16.99 ± 1.10
SGPT (U/L)	13.52 ± 1.29	18.63 ± 1.48	18.00 ± 1.10	18.15 ± 1.14	17.95 ± 0.83	17.90 ± 1.34
Alkaline phosphatase (U/L)	25.95 ± 1.89	42.47 ± 1.25*	26.78 ± 1.72$^{\#}$	30.65 ± 1.33$^{\#}$	30.65 ± 1.33$^{\#}$	32.77 ± 1.25*#
Leptin (ng/dl)	54.35 ± 1.49	48.51 ± 1.53	36.34 ± 1.32$^{*\#}$	29.63 ± 1.32$^{*\#\$}$	42.87 ± 1.32*	33.87 ± 1.62$^{*\#@}$
Adiponectin (μg /dl)	8.56 ± 0.34	8.68 ± 0.32	7.56 ± 0.23	8.18 ± 0.19	8.46 ± 0.23	8.75 ± 0.27

**Significant when compared to control.*
#Significant when compared to obestatin.
\$Significant when compared to GW9662.
@Significant when compared toT0070907.

Table 3.
Effect of obestatin (160 nmol/kg/BW), GW9662 (1 mg/kg/BW), T0070907 (1 mg/kg/BW), individually and in combination of obestatin + GW9662 (160 nmol/kg/BW + 1 mg/kg/BW, obestatin + T0070907 (160 nmol/kg/BW + 1 mg/kg/BW) treatment on fasting plasma biochemical parameters. Data are expressed as the mean ± SEM (N≧8). P < 0.05 was considered as statistically significant value.

DOI: http://dx.doi.org/10.5772/intechopen.103700

Plasma lipid parameters (mg/g)	Control	Obestatin	GW9662	Obestatin + GW9662	T0070907	Obestatin + T0070907
Epididymal fat						
Triglycerides	116.40 ± 1.37	108.05 ± 2.40	105.00 ± 1.77*	97.01 ± 2.02*#	107.00 ± 2.33	93.87 ± 2.56*#@
Total cholesterol	4.46 ± 0.13	4.86 ± 0.10	4.5 ± 0.11	4.12 ± 0.06#	3.98 ± 0.06*#	4.10 ± 0.07#
Phospholipids	0.46 ± 0.02	0.86 ± 0.01*	0.86 ± 0.02*	0.93 ± 0.1*	0.89 ± 0.03*	0.94 ± 0.02*
Inguinal fat						
Triglycerides	107.90 ± 2.11	120.5 ± 2.72	120.2 ± 3.27	110.9 ± 3.01	107.60 ± 3.88	103.00 ± 3.23@
Total cholesterol	5.03 ± 0.17	4.78 ± 0.07	4.67. ± 0.17	3.99 ± 0.22*#	4.12 ± 0.23*	3.87 ± 0.16*#
Phospholipids	0.60 ± 0.1	0.58 ± 0.07	0.45 ± 0.07	0.76 ± 0.05	0.55 ± 0.05	0.72 ± 0.09
Liver						
Triglycerides	32.68 ± 2.17	47.09 ± 3.38*	33.02 ± 2.33	39.66 ± 1.75	39.56 ± 1.75	42.66 ± 2.25
Total cholesterol	4.58 ± 0.28	5.26 ± 0.16	4.56 ± 0.22	4.45 ± 0.11	5.34 ± 0.21	5.54 ± 0.27
Phospholipids	7.73 ± 0.28	8.84 ± 0.11	8.15 ± 0.28	8.85 ± 0.33	7.87 ± 0.27	8.22 ± 0.28

*Significant when compared to control.
#Significant when compared to obestatin.
@Significant when compared toT0070907.

Table 4.
Effect of obestatin (160 nmol /kg/BW), GW9662 (1 mg/kg/BW), T0070907 (1 mg/kg/BW), individually and in combination of obestatin + GW9662 (160 nmol/kg/BW + 1 mg/kg/BW, obestatin + T0070907 (160 nmol/kg/BW + 1 mg/kg/BW) treatment on lipid parameters such as triglycerides, total cholesterol, phospholipid content in epididymal fat and liver tissue. Data are expressed as the mean ± SEM (N≧8). P < 0.05 was considered as statistically significant value.

method with GAPDH as the internal control gene. The experiment was done in "triplicates of triplicate" three experimental samples were estimated in triplicate to obtain statistical significance.

Figure 4 shows the mRNA profiling of the epididymal adipose tissue of mice treated with the respective compounds. In the epididymal adipose tissue, transcription factors PPARg, CEBPa, and CHREBP decrease by about 0.5 fold in obestatin, GW9662 and T0070907 treated groups whereas, obestatin + GW9662 and obestatin + T0070907 showed PPARg levels equal to that of the HFD control. CEBPa showed a 0.4 fold increase in both obestatin + GW9662 and obestatin + T0070907 treated groups. Concerning CHREBP, the obestatin treated group showed a decrease of 0.5% whereas, obestatin + GW9662 and obestatin + T0070907 groups showed 1.8 fold and 2 fold increase followed by GW9662 and T0070907 treated groups showing an increase by 0.4 fold above the control group.

Lipolysis was significantly enhanced in the obestatin + GW9662 and obestatin + T0070907 treated groups as seen from upregulation of ATGL and HSL by 0.3 fold and 1.8 fold respectively. GW9662 and T0070907 treated groups showed a 0.4% decrease in ATGL levels and no changes in HSL levels. Obestatin showed a decrease in 0.4 fold concerning ATGL and 0.4 fold decrease in HSL levels. No significant changes were seen to MGL.

No changes in mRNA levels of Perilipin 1 were seen in the GW9662 and obestatin + GW9662 treated groups. Whereas obestatin showed a 0.5 fold decrease in Perilipin levels and T0070907 and obestatin + T0070907 showed a decrease of 0.2 and 0.3 folds respectively.

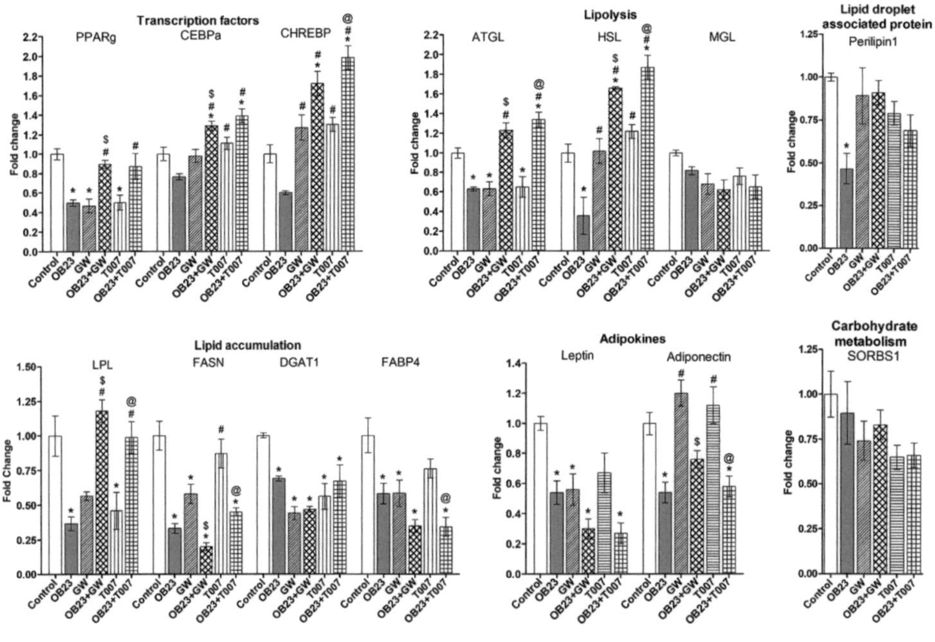

Figure 4.
Effect of obestatin (160 nmol /kg/BW), GW9662 (1 mg/kg/BW), T0070907 (1 mg/kg/BW), individually and in combination of obestatin + GW9662 (160 nmol/kg/BW + 1 mg/kg/BW, obestatin + T0070907 (160 nmol/kg/ BW + 1 mg/kg/BW) treatment on gene expression in epididymal tissue of obese C57BL/6 mice. β-Actin was used as internal control for gene expression studies by RT-PCR. Data are expressed as the mean ± SEM (N ≧ 8). P < 0.05 was considered as statistically significant value (= significant when compared to control, # = significant when compared to obestatin, $ = significant when compared to GW9662, @ = significant when compared to T0070907).*

With regards to lipid accumulation genes, obestatin + GW9662 showed an increase of LPL by 0.2 fold and obestatin + T0070907 showed no change with respect to the control. The obestatin, GW9662, and T0070907 treated groups showed a significant decrease by 0.7, 0.5, and 0.6 fold respectively. FASN was decreased in all treatment groups 0.8 fold by obestatin + GW9662 followed by obestatin by 0.7 folds, obestatin + T0070907 by 0.5 fold, GW9662 by 0.4 fold, and T0070907 by 0.2 fold. All treated groups showed decreased DGAT1 levels by 0.3–0.5 folds and FABP4 by 0.3–0.6 fold.

The adipokines leptin and adiponectin also showed differential regulation supporting a decrease in total fat content. Leptin levels were decreased by 0.7 folds by obestatin + GW9662 and obestatin + T0070907 groups whereas, obestatin and GW9662 showed leptin levels decreased by 0.5 fold. T0070907 showed 0.3 fold decrease in leptin levels. GW9662 and T0070907 groups showed 0.2 and 0.1 fold increase in adiponectin levels obestatin, obestatin + T0070907 showed a decrease in adiponectin levels by 0.5 fold and obestatin + GW9662 group by 0.3 fold. All treated groups showed a decrease in SORBS1 levels. T0070907 and obestatin + T0070907 at 0.4 fold, GW9962, obestatin + GW9662 and obestatin showed a decrease in SORBS1 levels by 0.3, 0.2 and 0.1 fold respectively.

Figure 5 shows the effect of the treatments on the inguinal adipose tissue. In inguinal adipose tissue, obestatin showed a 0.4 fold increase in PPARg and CHREBP levels and a 0.3 fold increase in CEBPa levels. All other treated groups showed a decrease in PPARg levels. GW9662 and obestatin + GW9662 showed a 0.4 fold decrease in PPARg levels. T0070907 and obestatin + T0070907 showed a decrease in PPARg levels by 0.9 and 0.6 folds respectively. A very similar trend was seen for

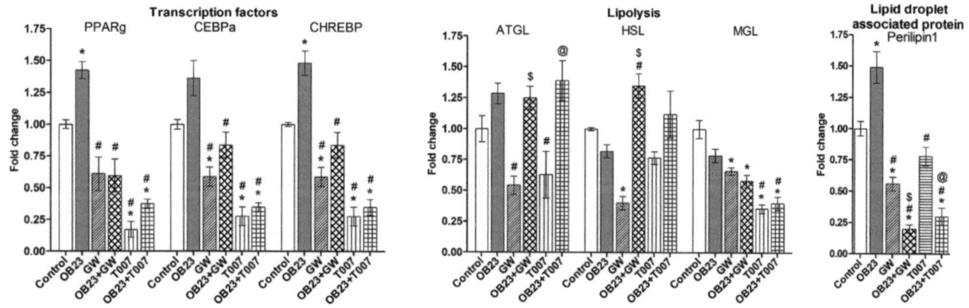

Figure 5.
Effect of obestatin (160 nmol /kg/BW), GW9662 (1 mg/kg/BW), T0070907 (1 mg/kg/BW), individually and in combination of obestatin + GW9662 (160 nmol/kg/BW + 1 mg/kg/BW, obestatin + T0070907 (160 nmol/kg/ BW + 1 mg/kg/BW) treatment on gene expression in inguinal tissue of obese C57BL/6 mice. β-Actin was used as internal control for gene expression studies by RT-PCR. Data are expressed as the mean ± SEM (N ≧ 8). P < 0.05 was considered as statistically significant value (= significant when compared to control, # = significant when compared to obestatin, $ = significant when compared to GW9662, @ = significant when compared toT0070907).*

all the treated groups except obestatin with respect to CEBPa and CHREBP with obestatin + GW9662 group showing a decrease by 0.2 fold, GW9662 by 0.4 fold, T0070907 by 0.75 fold, and obestatin + T0070907 by 0.7 fold.

With regards to lipolytic genes obestatin, obestatin + GW9662 and obestatin + T0070907 increased ATGL by 0.3 fold and GW9662 and T0070907 decreased by 0.5 fold with respect to the control group. Obestatin + GW9662 and obestatin + T0070907 increased HSL levels by 0.3 and 0.1 fold whereas obestatin, GW9662, and T0070907 decreased HSL levels by 02. 0.6 and 0.25 fold respectively. MGL was decreased by 0.7 fold by T0070907 and obestatin + T0070907, 0.4 fold by GW9662 and obestatin and GW9662, and 0.2 fold by obestatin.

Obestatin + GW9662 and obestatin + T0070907 showed decrease in Perilipin levels by 0.7 fold, T0070907 by 0.25 fold, GW9662 by 0.5 fold, obestatin + GW9662 by 0.8 fold. Whereas obestatin showed an increase in Perilipin levels by 0.5 fold.

LPL was decreased by 0.4 fold in GW9662, T0070907, and obestatin + GW9662 groups. T0070907 decreased LPL levels by 0.5 fold and the obestatin group showed an increase in LPL levels by 0.25 fold.

FASN was increased by obestatin by 0.75 fold whereas, all the other treated groups showed a decrease in FASN levels of 0.3 fold with exception of T0070907 at 0.7 fold. All treated groups showed a decrease in DGAT1. Obestatin, obestatin + GW9662, and T0070907 showed a decrease in DGAT1 levels at 0.5 fold, GW9662 at 0.25 fold, and obestatin + T0070907 at 0.75 fold.

No changes in FABP4 levels were observed in the obestatin and obestatin + GW9662 treated groups with respect to the control. GW9662 and T0070907 showed a decrease in FABP4 by 0.7 fold and Obestatin + T0070907 showed an increase in FABP4 levels by 0.25 fold.

Leptin was increased two-fold in the obestatin treated group, GW9662 and T0070907 showed a decrease in leptin levels by 0.4 fold. Adiponectin was decreased by the T0070907 and obestatin + T0070907 groups by 0.6 fold, GW9662, and obestatin + GW9662 groups by 0.25 fold and obestatin increased adiponectin by 0.2 fold.

SORBS1 levels remained unchanged in T0070907 and obestatin + T0070907 groups, GW9662 and obestatin + GW9662 groups decreased SORBS1 levels by 0.3 fold and obestatin showed no changes.

As seen in **Figure 6**, PPARg was enhanced by obestatin by 2 fold whereas, GW9662 and T0070907 decreased PPARg levels by 0.5 fold. Obestatin + GW9662 and obestatin + T0070907 decreased PPARg by 0.1 and 0.2 fold respectively.

FASN levels were decreased by 0.2 fold by all the treated groups except obestatin + T0070907, which decreased FASN levels by 0.6 fold.

GW9662 and obestatin + GW9662 groups increased SCD1 levels by 0.2 fold whereas, obestatin and obestatin + T0070907 showed no changes. The T0070907 group showed a decrease in SCD1 by 0.2 fold.

DGAT1 was decreased in all treated groups except obestatin. All the groups decreased DGAT1 by 0.5 fold and obestatin increased DGAT1 by 0.3 fold.

In brown adipose tissue, PPARg levels were enhanced by .75 fold by obestatin and obestatin + T0070907 and Obestatin + GW9662 and T0070907 by 1.5 fold. GW9662 enhanced PPARg twofold.

UCP1, responsible for thermogenic lipolysis of fat was increased by obestatin by 2.8 fold, obestatin + GW9662 by 1.6 folds obestatin + T0070907 by 1.5 fold. GW9662 and T0070907 increased UCP-1 by 0.1 and 0.3 fold.

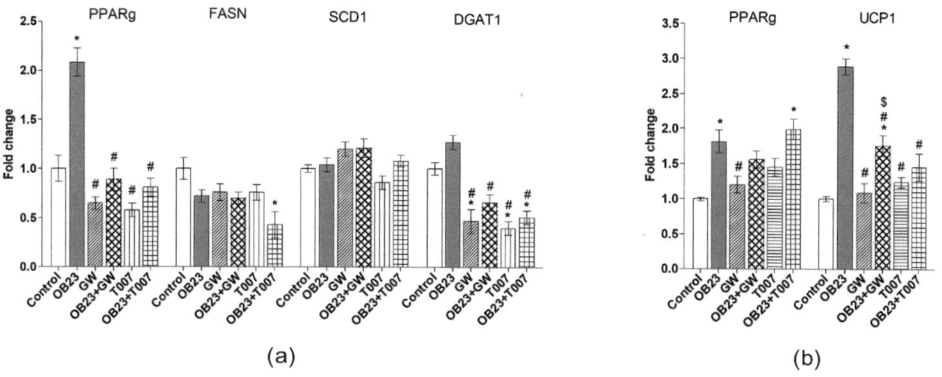

Figure 6.
Effect of obestatin (160 nmol/kg/BW), GW9662 (1 mg/kg/BW), T0070907 (1 mg/kg/BW), individually and in combination of obestatin + GW9662 (160 nmol/kg/BW + 1 mg/kg/BW, obestatin + T0070907 (160 nmol/kg/BW + 1 mg/kg/BW) treatment on gene expression in A. Liver tissue B. BAT of obese C57BL/6 mice. β-Actin was used as internal control for gene expression studies by RT-PCR. Data are expressed as the mean ± SEM (N ≧ 8). P < 0.05 was considered as statistically significant value (= significant when compared to control, # = significant when compared to obestatin, $ = significant when compared to GW9662, @ = significant when compared toT0070907).*

2.6 Comparison of the effect on lipid parameters observed upon the individual and coadministration of antagonists of PPARg GW9662 and T0070907 with rosiglitazone a well-studied agonist of PPARg

Administration of GW9662 and T00907 have shown the differential effect on plasma, adipose depots, and liver lipid parameters. All the treatments have shown a significant decrease in gain in body weight when compared to the HFD control. There is a decrease in food intake until 5 h after intraperitoneal administration of the agonists and obestatin, though there is no difference in the food intake after 12 h after administration. Both coadministrations show a decrease in inguinal and subcutaneous normalized fat pad weight. Whereas, triglyceride content is decreased in epididymal adipose tissue and total cholesterol in inguinal adipose tissue. The coadministration has significantly decreased plasma triglyceride and free fatty acids levels. Storage of fat in adipose tissue taking it out of circulation in the plasma is an indication of reduced lipotoxicity. Significantly decreased leptin levels in all the treated groups indicate the long-term effect of these compounds on lipid storage. mRNA profiling of the obestatin + GW9662 and obestatin + T0070907 groups show no change in PPARg levels, the significant increase in CEBPa and CHREBP levels indicating they do play a definite role in glucose and lipid metabolism in the liver and also adipogenesis [17–19]. An increase in ATGL and HSL levels also indicate the breakdown of triglyceride in the tissue and reduced perilipin1 levels indicate decreased storage of fat in the epididymal adipose tissue [20–22]. Decreased levels of FASN, DGAT1, and FABP4 are also indicative of decreased lipid accumulation [23–25]. Decreases in leptin levels seen in the plasma are also reflected at the mRNA level [26]. SORBS1 is a major regulator of insulin-stimulated signaling and regulation of glucose uptake, by potentiating insulin-induced phosphorylation and recruitment of CBL to a lipid raft [27]. A decrease in SORBS1 mRNA levels indicates altered glucose uptake in the epididymal adipose tissue.

On the other hand, in the inguinal adipose tissue, a decrease in the levels of transcription factors PPARg, CEBPa, and CHERBP in the antagonists and coadministration groups indicates decreased triglyceride accumulation. An increase in ATGL, HSL levels, and a decrease in MGL levels indicate decreased fat accumulation also supported by a decrease in Perilipin1 levels. Decreased lipid accumulation is also indicated by a decrease in the mRNA levels of FASN, DGAT1, and leptin. No changes are seen in SERBS1 levels indicating status co with insulin sensitivity.

In comparison, the PPARg agonist rosiglitazone showed an increased gain in body weight. Whereas, obestatin and obestatin + rosiglitazone showed a decrease in gain in body weight. Rosiglitazone showed significant a decrease in plasma triglycerides and free fatty acids by 50.93%, and 24.98% respectively. The same decrease was retained upon coadministration with obestatin. Rosiglitazone showed an increase in gluteal, cervical and subcutaneous fat and total fat content by 60%, 17.8%, 12%, and 20% respectively. Combined administration of obestatin and rosiglitazone reduced all the rosiglitazone increased fat content of gluteal, cervical, subcutaneous, and total fat to the control group levels [9]. In the epididymal adipose tissue, rosiglitazone showed a significant increase in LPL and FASN levels by 4.18 and 4.08 folds respectively. Whereas in the obestatin + rosiglitazone treated groups ATGL, HSL and MGL were upregulated by 3.78, 1.3, and 2.09 folds supporting a decrease in fat accumulation upon coadministration countering the fat accumulation increasing the effect of rosiglitazone. In Inguinal adipose tissue, obestatin + rosiglitazone upregulated ATGL by 2.83 fold. Leptin mRNA levels were upregulated 2.5 fold and 2.7 fold by rosiglitazone and obestatin + rosiglitazone [28].

3. Conclusions

The coadministration studies, previously with the PPARg agonist rosiglitazone and currently with PPARg antagonists GW9662 and T0070907 indicate that obestatin and PPARg agonists/antagonists show combined beneficial effects on plasma, liver, and adipose tissue lipid parameters. Obestatin reversed the lipid accumulation effect inherent to rosiglitazone. Whereas, the antagonists do not show a tendency towards lipid accumulation and additionally bring about beneficial effects by decreasing plasma and epididymal adipose tissue triglyceride levels. These studies indicate antagonists of PPARg in combination with obestatin could be furthered as lead compounds to counter obesity.

Acknowledgements

Mallikarjun BG gratefully acknowledges the financial support provided by ICMR for SRF (No.3/1/2/20/Nut./2012).

We acknowledge funding by The Council of Scientific and Industrial Research, India, through the major laboratory project MLP-116 and WELFO under the 12th five-year plan.

Abbreviations

ATGL	adipose triglyceride lipase
CEBPa	CCAAT/enhancer-binding protein alpha
CHREBP	carbohydrate-response element-binding protein
DGAT1	diacylglycerol O-acyltransferase-1

DIO	diet-induced obesity
FABP4	fatty acid-binding protein-4
FASN	fatty acid synthase
HSL	hormone sensitive lipase
PLIN	perilipin-1
PPARg	peroxisome proliferator-activated receptor gamma
SORBS1	sorbin and SH3 domain-containing protein 1
TC	total cholesterol
TG	triglyceride

Author details

Beekanahalli G. Mallikarjuna and Uma V. Manjappara*
Department of Lipid Science, CSIR-Central Food Technological Research Institute, Mysore, Karnataka, India

*Address all correspondence to: umamanjappara@cftri.res.in

References

[1] Zhang JV, Ren P-G, Avsian-Kretchmer O, et al. Obestatin, a peptide encoded by the ghrelin gene, opposes ghrelin's effects on food intake. Science. 2005;**310**:996-999

[2] Lauwers E, Landuyt B, Arckens L, Schoofs L, Luyten W. Obestatin does not activate orphan G protein-coupled receptor GPR39. Biochemical and Biophysical Research Communications. 2006;**351**:21-25

[3] Cowan E, Burch KJ, Green BD, Grieve DJ. Obestatin as a key regulator of metabolism and cardiovascular function with emerging therapeutic potential for diabetes. British Journal of Pharmacology. 2016;**173**(14):2165-2181

[4] Nagaraj S, Peddha MS, Manjappara UV. Fragments of obestatin as modulators of feed intake, circulating lipids and stored fat. Biochemical and Biophysical Research Communications. 2008;**366**:731-737

[5] Nagaraj S, Peddha MS, Manjappara UV. Fragment analogs as better mimics of obestatin. Regulatory Peptides. 2009;**158**:143-148

[6] Nagaraj S, Raghavan AV, Rao SN, Manjappara UV. Obestatin and Nt8U influence glycerolipid metabolism and PPAR gamma signaling in mice. The International Journal of Biochemistry & Cell Biology. 2014;**53**:414-422

[7] Reddy MSK, Manjappara UV. Capsaicin but not genistein influences modulation of lipid parameters by obestatin in DIO-C57BL/6 mice. International Journal of Peptide Research and Therapeutics. 2020;**26**:11-19

[8] Singh GR, Manjappara UV. Impromptu effects of Nt8U with soyasaponins on Obesity-related Lipid Parameters in high fat fed C57BL/6 mice. International Journal of Peptide Research and Therapeutics. 2020;**26**:405-411

[9] Mallikarjuna BG, Manjappara UV. Co-administration with Obestatin reduces accumulation of subcutaneous fat due to rosiglitazone administration in DIOC57BL/6 mice. International Journal of Peptide Research and Therapeutics. 2020;**26**:2307-2315

[10] Leesnitzer LM, Parks DJ, Bledsoe RK, Cobb JE, et al. Functional consequences of cysteine modification in the ligand binding sites of peroxisome proliferator activated receptors by GW9662. Biochemistry. 2002;**41**:6640-6650

[11] Lee G, Elwood F, McNally J, Weiszmann J, et al. T0070907, a selective ligand for peroxisome proliferator-activated receptor g. Functions as an Antagonist of Biochemical and Cellular Activities. 2002;**277**:19649-19657

[12] Nagaraj S, Manjappara UV. Studies on the influence of CCK-8 on the ability of obestatin to reduce food intake, gain in body weight and related lipid parameters. Biochimie. 2016;**125**:126-130

[13] Stewart JC. Colorimetric determination of phospholipids with ammonium ferrothiocyanate. Analytical Biochemistry. 1980;**104**:10-14

[14] Folch J, Lees M, Sloane Steyanly GH. A simple method for isolation and purification of total lipids from animal tissues. The Journal of Biological Chemistry. 1969;**4**:273-279

[15] Fletcher MJ. A colorimetric method for estimating serum triglycerides. Clinica Chimica Acta. 1968;**22**:393-397

[16] Searcy RL, Bergquist LM. A new colour reaction for the quantitation of serum cholesterol. Clinica Chimica Acta. 1960;**5**:192-199

[17] Spiegelman BM. PPAR-gamma: Adipogenic regulator and thiazolidinedione receptor. Diabetes. 1998;**47**:507-514

[18] Olofsson LE, Orho-Melander M, William-Olsson L, Sjöholm K, Sjöström L, Groop L, et al. CCAAT/enhancer binding protein alpha (C/EBPalpha) in adipose tissue regulates genes in lipid and glucose metabolism and a genetic variation in C/EBPalpha is associated with serum levels of triglycerides. The Journal of Clinical Endocrinology & Metabolism. 2009;**93**(12):4880-4886

[19] Ortega-Prieto P, Postic C. Carbohydrate sensing through the transcription factor ChREBP. Frontiers in Genetics. 2019;**10**:1-9. Article 472

[20] Kienesberger PC, Oberer M, Lass A, Zechner R. Mammalian patatin domain containing proteins: A family with diverse lipolytic activities involved in multiple biological functions. Journal of Lipid Research. 2009;**50**(suppl):S63-S68. DOI: 10.1194/jlr.R800082-JLR200

[21] Kraemer FB, Shen WJ. Hormone-sensitive lipase: Control of intracellular tri-(di-)acylglycerol and cholesteryl ester hydrolysis. Journal of Lipid Research. 2002;**43**(10):1585-1594

[22] Brasaemle DL, Subramanian V, Garcia A, Marcinkiewicz A, Rothenberg A. Perilipin a and the control of triacylglycerol metabolism. Molecular and Cellular Biochemistry. 2009;**326**(1-2):15-21

[23] Smith S, Witkowski A, Joshi AK. Structural and functional organization of the animal fatty acid synthase. Progress in Lipid Research. 2003;**42**(4):289-317

[24] Yen CL, Stone SJ, Koliwad S, Harris C, Farese RV. Thematic review series: Glycerolipids. DGAT enzymes and triacylglycerol biosynthesis. Journal of Lipid Research. 2008;**49**(11):2283-2301

[25] Furuhashi M, Saitoh S, Shimamoto K, Miura T. Fatty acid-binding protein 4 (FABP4): Pathophysiological insights and potent clinical biomarker of metabolic and cardiovascular diseases. Clinical Medicine Insights: Cardiology. 2014;**8**(Suppl. 3):23-33

[26] Brennan AM, Mantzoros CS. Drug insight: The role of leptin in human physiology and pathophysiology—Emerging clinical applications. Nature Clinical Practice.Endocrinology & Metabolism. 2006;**2**(6):318-327

[27] Baumann CA, Ribon V, Kanzaki M, Thurmond DC, Mora S, Shigematsu S, et al. CAP defines a second signalling pathway required for insulin-stimulated glucose transport. Nature. 2000;**407**(6801):202-207

[28] Manjappara UV. Thesis title Insights into the Mode of Action of Obestatin: The Role of PPARg. CFTRI: Mysore; 2020

Section 3

A New Approach to the Study of Peroxisomes

Chapter 7

Computational Methods for the Study of Peroxisomes in Health and Disease

Naomi van Wijk and Michal Linial

Abstract

Peroxisomal dysfunction has been linked to severe human metabolic disorders but is also linked to human diseases, including obesity, neurodegeneration, age-related diseases, and cancer. As such, peroxisome research has significantly increased in recent years. In parallel, advances in computational methods and data processing analysis may now be used to approach unanswered questions on peroxisome regulation, mechanism, function, and biogenesis in the context of healthy and pathological phenotypes. Here, we intend to provide an overview of advanced computational methods for the analysis of imaging data, protein structure modeling, proteomics, and genomics. We provide a concise background on these approaches, with specific and relevant examples. This chapter may serve as a broad resource for the current status of technological advances, and an introduction to computational methods for peroxisome research.

Keywords: high-resolution microscopy, structure prediction, integrative modeling, deep learning, genomics, proteomics, mass spectrometry, Zellweger syndrome

1. Introduction

Peroxisomes are single membrane-bound organelles found in all eukaryotic cells, with diverse functions according to the cell type and metabolic conditions. The study of peroxisomes, their processes, and regulation activity has taken flight in the last decade, thanks to the introduction of novel cellular methodologies. Simultaneously, sequencing and human genetics methods have strongly improved, and a growing number of metabolic diseases have been associated with genetic variations in peroxisomal genes. However, a deep understanding of the coordinated workings of peroxisomal genes in health and disease is still lacking.

Peroxisomes play a key role in cell metabolism and homeostasis. For example, they are involved in the β-oxidation of fatty acids, the formation of specific ether phospholipids, and in the dissipation of damage caused by reactive oxygen species (ROS). At the same time, peroxisomes are very dynamic organelles that use creative solutions in their biogenesis and assembly, import of large proteins, morphological changes by fusion and fission, and complex interactions with other organelles and lipid droplets. Several insightful review articles summarize the current challenges of the metabolism and biology of peroxisomes [1, 2]. This chapter explores the use of state-of-the-art computational approaches and methods for the study of peroxisome biology at various levels. First, we discuss the basics of deep learning and machine learning algorithms. Then, we explore how deep-learning-augmented cell imaging explores peroxisomal biology. Advances in microscopy have yielded a trove of high-resolution, high-throughput imaging data from living cells that require and enable a more advanced level of analysis. We then zoom in to a higher-level resolution, focusing on structural analysis and integrative modeling of peroxisomal proteins and their protein complexes. Next, we explore high throughout methods to study the interactions between proteins and lipids at the level of the proteome and lipidome. Finally, we discuss genetic approaches considering peroxisomal dysfunction and pathology. We highlight the role of computational and bioinformatics-based approaches to major open questions in the field of peroxisome biology.

2. Analysis of cell biology and imaging data

Observation remains one of the central pillars of biological research and cell biology in particular. Throughout the second half of the 20th century, electron microscopy (EM) images were fundamental to unveil the intracellular structures and organelles of a static cell at nanoscale resolution. The field of cell biology was revived by merging molecular biology techniques with *in vivo* dyes in the form of reporters and biosensors (e.g., GFP and its derivatives). The use of such fluorescence probes allowed monitoring cell dynamics under varying conditions while remaining as close-to-native state.

Current microscopy technology enables imaging at exceptional resolution, in the xy-plane varying from about 200 nm in confocal microscopy to about 10 nm in single-molecule localization microscopy [3]. Since peroxisome size ranges from 100 to 1000 nm in size, high-resolution microscopy is required for accurate and detailed imaging of protein expression and cellular localization. With the advances in high-resolution microscopy linked to automated, robotic lab support, the experimental results produce unprecedented big data of images and videos of cell dynamics. Although each technique has its advantages and limitations, the level of detail they reveal requires advanced data processing, analysis, and storage solutions. In past years, advanced algorithms, and especially deep-learning algorithms, developed for application to the biological domain, have skyrocketed.

2.1 Basics of deep learning

Supervised deep-learning algorithms require an annotated dataset, training on that dataset, and the use of the trained model on unseen, new data. In imaging applications, a dataset can be augmented and diversified by providing the model with edited (e.g., zoomed or rotated) images. The model is then trained on the dataset. The classification task aims to construct a function that takes this array as input to predict a label. In neural networks, which are usually used for classification problems, the learning task then aims to minimize the "loss function" (i.e., error), by optimizing a set of parameters, or weights, that multiply input data to obtain the output data that is passed on to the next layer in the neural network. A common loss function is cross-entropy to measure the difference between a true label and the predicted one. The algorithm architecture should be chosen to minimize overfitting, that is, when the model performs better on the training data than on a validation dataset. In the case of underfitting, the model performs poorer on the training set, resulting in suboptimal performance. For an elaborated review of the use of deep learning for cellular image analysis see ref. [4] and references within.

2.2 Advanced image analysis

The field of cellular and molecular biology had benefited from the availability of analytical tools for classical microscopy images. Major tools include CellProfiler, Microscopy Image Browser, OMERO, Fiji, and others [4]. Applying cell imaging tools according to unified standards led to the success of large-scale resources, such as the Human Proteome Atlas (HPA). HPA compiled enormous amounts of microscopical confocal images for annotating the subcellular information by using specific antibodies for most human coding genes across several human cell lines [5].

Recent years have seen the publication of updates of these tools with added components of deep-learning algorithms. Important applications of these tools include image classification, image segmentation, object tracking, and augmented microscopy. Specific examples of image analysis for the study of organelles are briefly discussed. **Figure 1** shows a scheme for addressing open questions in peroxisome biology using a large set of raw microscopic images (static and dynamic) and state-of-the-art methods of deep learning for the task of deciphering cross-organelle interactions [6, 7].

2.2.1 Image classification

In image classification, labels can be added to images, for example, to identify if a fluorophore-labeled protein is localized to a specific organelle or resides in the cytoplasm. For this purpose, machine-learning-based image classification has long focused on the generation of classifiers to identify changes in morphology following exposure to compounds or growth conditions or to identify changes in cell state.

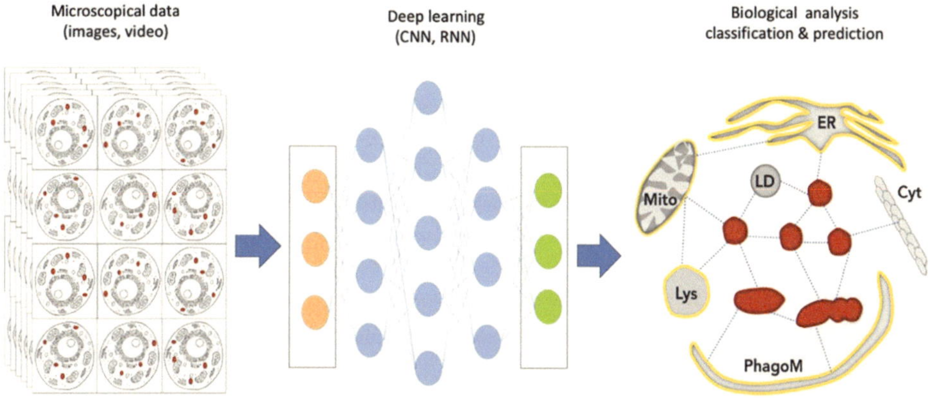

Figure 1.
A scheme for addressing open questions in peroxisome biology using a large set of raw microscopic images (static and dynamic). The image data is transformed by a convolutional neural network (CNN) whose output provides insights on large-scale cell biological challenges such as the task of organelles interactions, shape, and their dynamic crosstalk. Mito, mitochondrion; LD, lipid droplet: PhagoM, phagophore membrane; Lys, lysozyme, ER, endoplasmic reticulum; Cyt, polymerized cytoskeleton fiber (e.g., actin). Peroxisomes are colored red and membranes are colored yellow.

A popular software package called ilastik provides an addition to classical cell imaging tools with workflows for image segmentation, object classification, counting, and tracking. The pixel classification workflow produces semantic segmentation of images, attaching a user-defined class label to each pixel of the image. This step also forms the first step for object classification with morphological object features or may be used as initial input for the carving workflow. For example, ilastik was used for the reconstruction of 3D data from focused ion beam scanning EM (FIB-SEM) to segment the ER, using a pseudo-automated approach [8].

In a study by Li et al. [9], deep learning was used to classify organelle morphology of chloroplasts, mitochondria, and peroxisomes in the plant model Arabidopsis. The authors described a deep-learning framework, DeepLearnMOR (Deep Learning of the Morphology of Organelles), that identifies organelle morphology abnormalities at 97% accuracy. In the study, a dataset of 47,000 confocal fluorescence microscopy images from Arabidopsis wild-type and mutant plants with abnormal division in chloroplasts, mitochondria, or peroxisomes, was used to train the model. The dataset was augmented by using rotated, flipped, and split images. The model is based on both transfer learning and convoluted neural networks and significantly outperformed conventional machine-learning methods. In deep learning, transfer learning entails training a model on a large dataset and then fine-tuning the model for a different task using a new, smaller dataset. In this framework, the model distinguished well between mitochondria and peroxisomes, despite the overlap in their sizes. The framework can be used to study subtle morphological changes to classify intact and aberrant human peroxisome morphology.

In another example, a multi-scale convolutional neural network approach was developed and trained on eight publicly available cellular imaging datasets [10]. Following training, both the binary phenotype classification task as well as a multi-label classification task, performed at least as good as state-of-the-art architectures, saving time on the manual adjustment of parameters for segmentation and feature selection, that is needed for conventional image analysis pipelines. The datasets included images of stains for various organelles and cell types. Although peroxisomes were not included in the datasets used, this approach can be used for expanding the classification to cover peroxisome phenotypes.

An additional classic classification question regards the observation that proteins are localized into multiple subcellular compartments. About half of human proteins exist in more than one organelle simultaneously, and these multi-locational proteins are likely to play critical roles in cellular functions [5]. To deal with these multilabel proteins, most existing methods converted the multilabel classification problem into L binary problems, L being the number of classes. However legitimate, such simplified approaches ignored label dependencies that actually exist among subcellular locations [11]. The difficulty of the classification part mainly lies in the multilabel nature of protein localization, as is also exemplified in ref. [10]. Even in a simplified setting of a cell line in culture, cells are at different stages of cell division and density, yielding nonuniform localization profiles.

2.2.2 Image segmentation

Image segmentation is the task of identifying multiple objects or features within an image, for example, cell counting. LysoQuant is a deep-learning approach for the detection and segmentation of organelles and is available as an ImageJ plugin. Its efficacy was demonstrated using the ER as a model organelle and a polymerogenic α1-antitrypsin Z (ATZ) variant as a model disease-causing aberrant protein [12].

The model's performance was validated on the quantification of catabolic pathways that maintain cellular homeostasis and proteostasis. The model was tested on two cell types and on the ER as a model organelle, but it may very well be applicable for use in other cell types and for the study of peroxisomes.

Due to the advancement of classification accuracy and the availability of high computational power, cell image segmentation approaches are often based on deep convoluted neural networks. One of the disadvantages of deep-learning approaches is the large amount of training data required to train them. Although software packages, such as CellProfiler, already use deep-learning models, they do not support retraining on new data, thus restricting their application domain to an available set of datasets. In contrast, U-Net is pretrained on a diverse set of data and for every new task needs only a few (<10) annotated images [13].

In addition, sequential images often differ one from the other in the sense that the number of objects belonging to each class differs from image to image, leading to an imbalance in the class weights. One approach to tackle this problem is by automatically updating the weights of the imbalanced classes by constructing a new objective function. In one recent study, a U-Net-like convoluted neural networks model is used with two updated loss functions to improve segmentation of cell organelles, including cytoplasm, plastics, nucleus, mitochondrion, and peroxisome [14]. This demonstrates the importance of adapting the loss function in deep-learning approaches for improving the success of segmentation tasks.

2.2.3 Object tracking

The challenges in imaging described so far assume cells to be rather static objects. Obviously, the dynamics and heterogeneity of cells define their biology. In object tracking, objects are followed through a series of time-lapse images. This requires two tasks—object detection and object linkage. Single-particle tracking (SPT) is often the rate-limiting step in live-cell imaging studies of subcellular dynamics. Many object tracking models are based on a tracking algorithm that addresses the principal challenges of SPT, namely high particle density, particle motion heterogeneity, temporary disappearance of particles, and merging and splitting of particles. The algorithm first links particles between consecutive frames and then links the resulting track segments into complete trajectories [15]. This approach forms the basis for software such as CellProfiler and TrackMate.

TrackMate is a software that offers several detections and tracking modules that allow combining manual and automated particle tracking approaches. An openly available tool, it is available as an extension of ImageJ. Moreover, the capabilities of the software can be tailored by the user through the addition of specific tracking, detection, visualization, or analysis modules. Its data model makes it a useful tool for a wide range of tracking applications, ranging from single-particle tracking of subcellular organelles to cell lineage analysis. Importantly, the TrackMate study stresses the importance of avoiding photoinduced stress due to the continuous or repetitive illumination required for fluorescence microscopy [16].

An interesting example of the use of TrackMate for tracking, along with other image analysis methods, is presented in ref. [17]. While current imaging techniques are constrained by the small number of distinctive fluorescent labels within a single image, the use of confocal and lattice light sheet (LLS) fluorescence microscopy combined with computational sophistication allowed to track globular organelles and propose dynamic inter-organelle contacts. The study describes the frequency and locality of two- to five-way interactions among major membrane-bound organelles (ER, Golgi, lysosome, peroxisome, mitochondria, and lipid droplet) and shows how these relationships change over time.

2.2.4 Image augmentation

Augmented microscopy is the extraction of latent information from biological images, such as the identification of the locations of cellular nuclei in bright-field images. Many augmented microscopies approach train neural networks to translate between label-free (bright-field, phase, differential interference contrast, and transmission EM) and labeled (fluorescence) images of the same cells. The ability to predict fluorescence images from grayscale data is advantageous for increased imaging speed and improved time-lapse imaging. Moreover, this neural network methodology was adapted from the classical field of imaging processing and its implementation to organelle biology takes advantage of the overwhelming amounts of grayscale images already produced by standard biology labs [18]. In this study, the prediction performance across organelles and other subcellular structures reach an accuracy of detection between 70 and 90% for cellular compartments, such as nucleoli, nuclear envelope, mitochondria, and ER [18, 19]. Undoubtedly, the model can be trained and tested for the study of peroxisomes as well.

3. Protein structure modeling and analysis

Protein structure modeling and prediction are experiencing exciting times. The "protein folding problem," which has puzzled scientists for many decades, asks to predict a protein's structure from its primary amino acid sequence. It is thus not surprising that artificial intelligence (AI)-driven protein structure software, which includes both AlphaFold and RoseTTAFold, were announced as the Science breakthrough of the year 2021. In this chapter, we will discuss these models, followed by a discussion of how protein prediction models may be integrated into classical 3D experimental methods, such as cryo-EM, X-ray crystallography, mass spectrometry (MS), and nuclear magnetic resonance (NMR). We will then discuss approaches for modeling protein–protein interactions (PPI) and protein complexes.

3.1 AlphaFold

The Critical Assessment of protein Structure Prediction (CASP) was initiated in 1994 as a biennial open competition for advancing methods for 3D protein structure prediction. Until 2016, the average prediction accuracy score across multiple approaches was bound to 30–40 (on a scale of 100). In 2018, the AlphaFold model developed by the DeepMind company (owned by Google) scored ~55, and in 2020, AlphaFold2 reached an astonishing score of ~92. For reference, a score > 90 is roughly equivalent to the variation monitored from repeated experiments for determining protein structure [20].

The input of the AlphaFold model is based on the primary amino acid sequence of a protein and the sequence alignment compilation of known homologs. Specifically, the amino acid sequence is used to build a multiple sequence alignment (MSA) from similar sequences found in protein sequence databases. Assuming an accurate MSA, the amino acid pairs that were co-mutated along the evolution path can be detected. A crude structure representation, or "pair representation" is then produced based on structural templates and on the paired amino acids that are likely in contact with each other. The templates and MSA are then passed through a transformer termed "Evoformer" that takes the MSA representation and the pair representation, to refine these representations. In the final step, the "structure module" takes the refined MSA representation and pair representation to construct a refined 3D structure model of heavy atoms [21].

The predicted structures contain atomic coordinates and per-residue confidence estimates on a scale from 0 to 100, with higher scores corresponding to higher confidence. This per-residue confidence measure, pLDDT, is based on a preexisting metric used in the protein structure prediction field. Scores >70 are considered residues predicted with confidence and > 90 are considered as very high confidence prediction (**Figure 2**). It should be noted that in multi-domain predictions, individual domains may be largely accurate while their relative position is not [22, 23].

As of January 2022, AlphaFold database provides free access to >360,000 predicted structures across 21 proteomes. In comparison, the Protein Data Bank (PDB) that was announced 50 years ago contains "only" 180,000 experimentally solved 3D structures, and many of them do not cover the full length of the protein. However, the UniProtKB database includes almost 220 million entries. For many of the PEX proteins (peroxins) involved in peroxisome biogenesis, no experimentally determined structures are available [24]. However, a search for PEX proteins in the current AlphaFold database identified 28 hits from yeast, 26 from Arabidopsis, and 22 from human origin.

For example, the extensively studied peroxisomal targeting signal 1 receptor (PEX5), is represented in the PDB archive (nine entries) that mostly covers the well-folded domain at the C′ terminus. The AlphaFold database, however, includes a prediction of the full-length protein, which includes a very high confidence prediction for the C′ terminal domain, but also revealed some helices that were previously not solved (**Figure 2A**). It should be noted that the low-confidence regions in AlphaFold predictions largely correspond with known intrinsically disordered regions (IDRs) [25]. For the human peroxisomal targeting signal 2 receptor (PEX7) protein, no experimentally solved structures are available. However, the AlphaFold

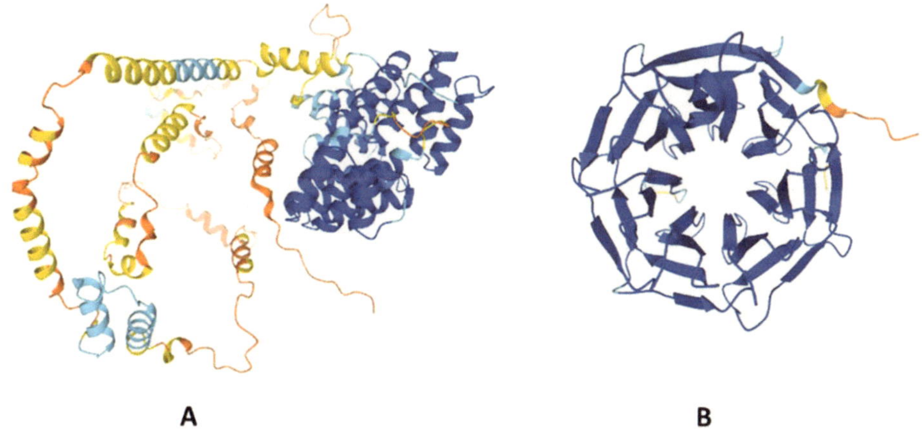

Figure 2.
Prediction of PEX5 (A) and PEX7 (B) structure by AlphaFold2. Model confidence—Dark blue, very high (pLDDT > 90); light blue, confident (90 > pLDDT > 70); yellow, low (70 > pLDDT > 50); orange, very low (pLDDT < 50). Note that many of the low-confidence structure predictions coincide with intrinsic disorder in the protein.

predicted protein structure is a strikingly beautiful β-propeller structure with sevenfold symmetry, similar to a published structure of its yeast homolog (**Figure 2B**) [26]. It is now possible to compare structural differences between the yeast and human homologs to shed light on differences in function. We anticipate that many structural and mechanistic questions regarding peroxisomal proteins can now be approached with a wealth of reliable modeled structural data.

3.2 Rosetta

The Rosetta software suite was initially developed in the Baker Lab at the University of Washington, Seattle, and was evolved as an active collaborative effort [27]. It includes many functionalities for macromolecular modeling. Its applications and protocols include *ab initio* modeling [28–30], including membrane protein modeling [31–33], comparative modeling using one or more known structures as templates for modeling, "fold-and-dock" application for the prediction of symmetric homo-oligomer structures, FlexPepDock for *ab initio* or refinement of peptide docking to a receptor, "relax" that improves protein energy landscape modeling and more.

Around the same time that the AlphaFold model was published following its winning entry at the CASP14 competition, RoseTTAFold, another extremely successful protein structure prediction model, was published. Like AlphaFold2, it uses deep neural networks to find sequence patterns in databases of similar sequences. When given a new sequence to model, RoseTTAFold proceeds along multiple tracks—one creates an MSA, another predicts pairwise interactions between amino acids within the protein, and the third constructs the 3D model structure. The program bounces among the tracks to refine the model, using the output of each one to update and refine the others [34, 35].

The trRosetta model is also Rosetta-based and incorporates restraints for prediction. The algorithm starts off with MSA for distance and contact prediction to learn probability distributions over distances between residues and determine residue orientation. The predicted distances and orientations are then used to generate 3D structures using constrained energy minimization. The lowest-energy backbone is then subjected to Rosetta full-atom relaxation to add side chains and make the structures physically plausible, and to generate the lowest-energy full-atom model. The trRosetta network is able to identify the most important residues for determining protein folding and can apply this on *de novo* designed proteins, although the model was only trained on native proteins [36].

3.3 Modeling of protein: protein interactions

The prediction of multi-chain protein complexes is an even greater challenge than the prediction of monomeric protein structures. In the context of peroxisome biology, the functional and organelle mysteries reside in the dynamic interaction between the membranal and matrix proteins [2]. In addition, PEX protein's subcellular localization is governed by post-translation modifications (PTMs), such as ubiquitination. All these aspects are executed by the dynamic of protein–protein interactions (PPI).

Many docking algorithms are being used for *ab initio* or template-based modeling, several of which we will discuss briefly. Some of the first methods for multimolecular modeling, developed in the Wolfson Lab at Tel Aviv University, include CombDock, PatchDock, FlexDock, and more recently, DockStar [37]. Input for the CombDock algorithm is an ordered set of protein sub-structures, which then combinatorically assembles the inter-contacts that define their overall organization. PatchDock and FlexDock are based on the integration of local shape complementary evidence, where both input molecules are considered rigid, or in which one of the input molecules is considered rigid and the other flexible. DockStar integrates both low-resolution information (e.g., MS) and high-resolution (e.g., X-ray, NMR, and homology modeling) experimental data, combining atomic structures and interaction data. It then calculates the optimal assembly of the individual subunits.

Another approach is used by pyDock, which uses electrostatics, desolvation energy, and in a limited manner, van der Waals forces, to score rigid-body docking poses. InterEvDock and InterEvDock2 use co-evolutionary information in docking based on rigid-body sampling. In InterEvDock2, protein sequences can be provided as input, not only 3D structures. The algorithm then first performs comparative modeling based on template search. If biological input is available such as a pair of residues known to be in contact, restraints with a tunable distance threshold can be specified for use in the docking procedure. A recent review of multi-molecular modeling approaches can be found in ref. [38].

The latest developments in AI-harnessed modeling approaches will likely become some of our most important tools for the modeling of PPI and protein complexes. AlphaFold-Multimer is an AlphaFold model trained specifically for multimeric inputs of known stoichiometry. For example, an A2B2C2 heteromer was solved with a structural prediction score of 98, virtually identical to the solved structure for this complex [39]. In addition, AlphaFold2 was shown to successfully predict peptide-protein complexes even though it was trained only on monomer chains [40]. Open questions regarding peroxisome complexes can now be approached using this updated model. For example, the human PEX2-PEX10-PEX12 proteins form a protein-ubiquitin ligase complex for which currently no structure is available. The same is true for the AAA+ ATPase heterotrimeric complex PEX1-PEX6, and more.

RoseTTAFold has comparable performance in identifying PPI to that of experimental methods, but the combination of applying the RoseTTAFold model with AlphaFold further increases identification accuracy. A combined protein interaction identification pipeline that incorporates a rapidly computable version of RoseTTAFold with the slower but more accurate AlphaFold, evaluates interactions between the 8.3 million possible pairs of yeast proteins. In total, 106 previously un-identified assemblies and 806 that were structurally uncharacterized, were modeled. These models include higher-order complexes up to pentameric assemblies [41]. This combined approach demonstrates the strength of combining various neural network-based models to maximize modeling accuracy and speed.

3.4 Integrative modeling and analysis

The algorithms described above in some cases use experimentally solved structures as templates for the modeling of protein complexes. These are instances of integrative structural modeling, which involves the determination of macromolecular structures by combining experimental and computational modeling approaches [42].

Workflows for experimental methods can be improved with modeling approaches at various steps. In X-ray crystallography, protein modeling can be used to improve the determination of the protein structure, which is hampered by the "phase problem" that prevents the direct determination of the 3D structure from X-ray diffraction data. This approach was used by combining AlphaFold modeling with

X-ray diffraction data to the determination of the structure of the ORF8 protein of SARS-CoV-2 [43].

Cryo-EM is more and more often used for structure determination. The highest resolution structures solved with cryo-EM have been solved at less than 2 Å [44]. This improvement in resolution has made cryo-EM a likely candidate to replace X-ray crystallography as the gold standard of experimental structure determination. A variety of software is available for modeling macromolecular assemblies using cryo-EM for *de novo* modeling, fitting, and validation of the atomic model. For high-resolution structures, data integration can be used for *de novo* model building and local refinement, whereas intermediate or low-resolution structures may benefit from rigid fitting and fully integrative modeling [45]. An impressive example of integrative modeling was the publication of the full structure of the 52 MDa nuclear pore complex (NPC) from yeast. In that study, protein modeling was combined with data obtained from cryo-electron tomography (cryo-ET), a method similar to cryo-EM, chemical cross-linking mass spectrometry, *in vivo* imaging, and existing solved structures of NPC subunits, as the experimental data input [46]. The successful use of *ab initio* modeling is further illustrated by the cryoSPARC software, which makes it possible to very quickly perform unsupervised *ab initio* 3D classification to discover multiple 3D states of a protein without prior structural knowledge [47]. CryoSPARC is an end-to-end solution for cryo-EM analysis and structure refinement by combining 3D reconstruction algorithms with specially designed software.

NMR may benefit from systematic back-calculation of expectation spectra across a conformational space should then allow reconstruction of the experimental spectra. This would enable the comparison of the back-calculated and experimental data and provide us with a quantitative quality measure [48]. In addition, complex NMR results can benefit from the strength of a deep neural network-based approach such as DEEP Picker, to aid in the analysis of NMR spectra and correctly characterize overlapping peaks.

In cross-linking MS (CL-MS), an *in vitro* or *in vivo* cross-linking reaction covalently binds specific residues. The cross-linked proteins are then enzymatically digested, followed by liquid chromatography–tandem mass spectrometry (LC–MS/MS). The covalently linked peptides provide valuable information about spatially close residues, whether intramolecular or intermolecular, which are then used as constraints in structural modeling [49]. XL-MS data can be useful in the validation and refinement of predicted structures as well. The benefit of this integrative modeling approach was recently demonstrated in the structural determination of three proteins from SARS-CoV-2. In this study, docking algorithms PatchDock, CombDock, and AlphaFold2 were used to model the proteins and refine them with CL-MS data that were collected from living cells. It was found that intradomain cross-links were satisfied in most cases, whereas interdomain cross-links were often violated, demonstrating the strength of this integrative modeling approach [50].

Importantly, the integration of diverse data sources into unified predictive models is likely to advance the knowledge of the protein complex of peroxisomes. While the structural study of peroxisomal proteins is in its infancy, we demonstrate the strength of recently developed computational software, tools, and algorithms to integrate data using breakthrough AI-based structural prediction approaches and integrative modeling.

4. Proteomics

Although the studies of individual proteins are important, they mostly ignore the PPI, cellular and physical environment of the studied protein. In recent years, proteomics approaches have become extremely valuable for the systematic analysis and discovery of the involvement of and interaction between proteins at the cellular or organellar level. Depending on the goal of the study, proteome analysis may include the determination of the full cellular or organellar proteome or PPI. The abundance profiles of proteins throughout all fractions of the purification can be compared to the profile of known marker protein to identify proteins that co-fractionate with the organelle of interest. To identify interactions, the abundance of interactors to a tagged protein of interest is compared to the abundance of the same, untagged protein. In addition, proteome dynamics can be followed by analyzing the organellar proteome at various time points following a stimulus [51]. Here, we discuss some of the most advanced methods for the analysis and application of proteomics data, with an emphasis on mass spectrometry (MS)-based proteomics. Importantly, MS and other advanced quantitative proteomics highlight regulation that cannot be explored using nucleic acids sequencing approaches. In the context of peroxisomal biology, the information includes the presence of protein variants, PPI, subcellular localization, and the status of post-translational modifications (PTMs). Analyzing protein levels in an organelle is fundamental to exploring molecular signaling and dynamics in response to varying conditions.

4.1 Mass spectrometry-based proteomics

For organelle proteomics, three approaches can be taken to provide input for MS analysis, namely data-dependent acquisition, data-independent acquisition, and targeted proteomics, each of which is briefly discussed below. Following separation of the organelle, the organelle fraction can be run on SDS/PAGE to separate proteins and provide more specific input samples according to the size range of protein bands. The protein bands are then digested by specific proteases (e.g., trypsin) in-gel and analyzed. Alternatively, the organelle can be solubilized as a whole and digested in solution. The digested product is then analyzed. In addition, the organelle fraction can be fed into a workflow of liquid chromatography coupled with tandem mass spectrometry (LC–MS/MS) and subsequent data analysis [52]. A number of MS-based proteomics studies were undertaken in recent years to yield a comprehensive list of the mammalian proteome from specific tissues and cell lines [53].

4.1.1 Data acquisition approaches

In data-dependent acquisition (DDA), the eluted peptides from LC are detected by the first MS step, usually within a wide range of a mass-to-charge ratio of 400–1,200 m/z. The most abundant peptides are then entered into the second MS step following fractionation to yield a more informative peak spectrum [54]. However, if the number of precursor ions exceeds the number of precursor selection cycles, peptides detected in repeat analysis become irreproducible [55]. An inherent difficulty in quantitative MS proteomics is that only a few peptides are consistently identified in a complex protein mixture. The uncontrolled complexity of biological samples leads to poor reproducibility of MS-identified peptides. Characteristic features of prototypic peptides and their physicochemical properties were the basis for developing successful computational tools for predicting peptides for any organism [56]. To improve the detectability of all peptides in a sample, in data-independent acquisition (DIA), the second MS step receives as input, not individual peaks, but instead all peaks within a defined m/z range, thus using all peptide precursors within a mass range of interest. Usually, most tryptic peptides are within the 400–1200 m/z range, so all the initial peaks within this range are further specified and analyzed [57]. Finally, in targeted proteomics, selected or multiple reaction monitoring (SRM/MRM) is applied, meaning that a set of key peptides from a target list is quantitatively followed in many samples. In this case, the mass spectrometer fragments and analyzes only those peptides, increasing sensitivity and enabling the study of low-copy number proteins [58].

4.1.2 Proteomics analysis platforms

Various proteomics workflow software and pipelines are available, including Proteome Discoverer, MaxQuant, Mascot, OpenMS, CompOmics, and ProteomicsDB, as described in ref. [59]. These tools have developed drastically over the years, including more and more deep learning and advanced statistics capabilities. For example, MaxQuant is an open-source software package that supports DDA data and has an integrated peptide search engine, Andromeda. From the same developers, Perseus provides statistical tools for high-dimensional omics data analysis covering normalization, pattern recognition, time-series analysis, cross-omics comparisons, and multiple-hypothesis testing. MaxDIA was recently added to the MaxQuant pipeline and provides deep learning-based analysis of DIA data. Similarly, OpenMS is a much-used open-source software framework that enables both analyses of proteomics and metabolomics data. As such, it may be an especially relevant tool for peroxisome research. In addition, Proteome Discoverer includes deep-learning algorithms for the construction of spectral libraries as well as for the improved analysis of low-quality MS/MS spectra.

The retention time of a peptide refers to the time it takes for a peptide to elute from a liquid chromatography (LC) column prior to analysis by MS. As such, a peptide's retention time is determined by the degree to which it interacts with the column, and as such is highly reproducible under the same LC conditions. The accurate prediction of peptide retention time can be used to improve the sensitivity of peptide identification against a peptide database. Early models such as SSRCalc are based on the retention times of 2000 peptides and base predictions on peptide sequence [60, 61]. More advanced, deep learning-based methods can be used for building libraries of MS spectra to enable data analysis from DIA MS. For example, the ProteomicsDB was recently expanded with Prosit, a neural network model that accurately learns and predicts chromatographic retention time and fragment ion intensity of any peptide, both tryptic (i.e., digested by trypsin) and non-tryptic. Other examples, based on different types of neural network algorithms, include DeepRT and, more recently, DeepDIA. The latter approach generates *in silico* spectral libraries, comparable to experimental libraries, to enhance DIA analysis. The various neural network-based models require large training datasets, often in the range of 100,000 peptides or more. However, the pretrained model can then be trained on much smaller experiment-specific datasets that make the model very accurate. The main challenges in this field remain retention time prediction of modified peptides and cross-linked peptides [62].

4.2 Proteome-wide characterization of peroxisome proteins

The determination of the peroxisome proteome, as well as the interaction networks between them and the modifications they undergo that affect their function, can be done using a variety of computational tools and approaches. The proteome-wide study of peroxisome proteins is especially interesting to understand regulatory and functional aspects. Early experiments identified nearly all peroxisomal proteins, but it was then still challenging to discriminate between genuine peroxisomal proteins and co-isolated non-peroxisomal proteins, mostly derived from interacting organelles (e.g., mitochondria). The increasing sensitivity of MS enabled the detection of rare peroxisomal proteins, using improved experimental procedures such as organellar profiling [63, 64]. A more recent study of the proteome of HeLa cells provides a proteomic workflow for the generation of reproducible organellar maps. In the study, organellar clusters were assigned proteins based on a machine learning approach [65]. Based on this and other studies, an updated list of the mammalian peroxisomal proteome was made available in ref. [53]. However, many proteins are found in multiple organelles and peroxisome composition may vary between cell types.

An alternative approach to the MS-based proteomics approach described above is the prediction of localization to the peroxisome using peroxisomal targeting signals [66]. Matrix proteins are directed to the peroxisome with a type I or type II peroxisomal target sequence (PTS1 and PTS2) and are transported following the binding of PEX5 or PEX5/PEX7, respectively. In most cases, membrane proteins contain either of two types of PTS (mPTS-I/mPTS-II) and are inserted into the membrane by PEX19 and PEX3. Prediction algorithms for mammalian PTS1 motifs were published almost two decades ago and include PTS1-predictor, PTS1Prowler, PeroxiP, and an algorithm included in the PeroxisomeDB database. PTS1-predictor uses a position-specific scoring matrix (PSSM) or position weight matrix (PWM) approach, derived from aligned sequences of proteins known to harbor a PTS1 sequence, but also peptide sequences of proteins bound to various PEX5 homologs. Thus, the tripeptide dataset used for PTS1 prediction includes a larger amount of variations to find less probable proteins as well [67, 68]. The prediction of PTS2 proteins is much more challenging due to the small number of proteins bearing this signal peptide. A PTS2 prediction tool with a limited success rate is included in PeroxisomeDB. Finally, the In-Pero pipeline, based on a machine learning approach, was recently published for the prediction of sub-peroxisomal cellular localization of unclassified peroxisomal proteins.

The local concentration of functionally interconnected proteins yields PPI, the physical contact between two proteins, which may occur in a binary manner or may exist in multimeric complexes. Several proteome-wide PPI maps are available, including HI-II-14, BioPlex 3.0, and CoFrac. The number of protein pairs in each of these maps is well above 10,000, but the overlap between the maps is very limited [69]. The HI-II-14 map includes only binary interactions, generated from yeast-2-hybrid assays, whereas the others are based on affinity purification or co-fractionation followed by MS. An interesting development is the combination of various databases and depositories, such as the recently published MuSIC 1.0 map that integrates immunofluorescence images from the Human Protein Atlas (HPA) with affinity purification data from BioPlex 2.0 [70].

In addition to the interaction between proteins, PTMs include about 300 types of modifications, including phosphorylation, ubiquitination, glycosylation, acetylation but also regulated peptide cleavage. The large number of potential PTMs strongly affects protein function. It is expected to increase the proteome's complexity by at least an order of magnitude. PTMs can be identified experimentally in high-throughput MS approaches or in small experiments. However, deep learning also enables accurate prediction of whether a given site can be modified (general site prediction) and if a site can be modified by a specific enzyme (enzyme-specific prediction). Multiple PTMs can be predicted using CapsNet or MusiteDeep. In addition, specific modifications can be accurately predicted with dedicated tools like DeepPhos for phosphorylation, DeepAcet for acetylation, DeepUbiquitylation for ubiquitylation, and more [62].

4.3 Metabolomics

Although not strictly a part of the proteomics field of study, we briefly discuss the understudied metabolomics methods as they are of especial importance in the context of peroxisome metabolism. Metabolomics studies are commonly done using NMR or, more frequently, MS for the analysis of whole-cell or subcellular fractions. As such, much of the metabolomics pipeline is similar to that of the proteomics pipeline—gas chromatography (GC) or liquid chromatography (LC) coupled with MS or tandem MS (MS/MS).

Metabolomics data analysis can be done in an unsupervised or supervised manner. The goal of the unsupervised analysis is the grouping of features (sample, metabolites, and spectral features) according to the measured molecular data, and as such, this approach is suitable when no prior information is available about the system. In contrast, in supervised analysis, a set of features is pre-assigned to a class and is used as a training set for the method of choice to define a classifier that will be used for the classification of an unknown sample [71]. In addition, metabolomics studies can be divided into untargeted and targeted studies. Untargeted, also referred to as discovery-based metabolomics, focuses on global detection and relative quantitation of small molecules in a sample. In contrast, targeted or validation-based metabolomics focuses on measuring well-defined groups of metabolites with opportunities for absolute quantitation [72]. Several metabolomics data analysis tools are available, one of the most popular being MetaboAnalyst. This web-based tool suite covers four analysis categories, including statistical analysis, functional analysis, data integration and systems biology, data processing, and utility functions.

One of the challenges of subcellular or organellar metabolomics is the difficulty to observe metabolic fluxes between compartments. Metabolic flux studies are usually done directly by using isotopically labeled nutrients and measuring isotopically labeled metabolites to infer flux via metabolic flux analysis (MFA) or flux balance analysis (FBA) [73]. In FBA, the flow of metabolites through a metabolic network is analyzed mathematically. Two widely used computational flux inference approaches include isotope tracing coupled with MFA (13C-MFA) and constraint-based reconstruction and analysis (COBRA) [74]. Spatial flux analysis was done for the mitochondria and cytosol using 13C-MFA: isotope tracing in intact cells and subsequent rapid fractionation and metabolism quenching, followed by LC–MS-based metabolomics. Computational deconvolution with metabolic and thermodynamic modeling was used to infer compartment-specific metabolic fluxes [75]. In COBRA, which integrates various experimental and -omics data sources to reconstruct metabolic networks, applying constraints, for example, mass conservation, maximum reaction rates, and regulation, to construct a space of allowed network states [76]. These approaches could be used to model differences in metabolic flux in healthy and mutated peroxisome factors.

5. Human genetic research and peroxisomal disease

The contribution of peroxisomes to cells and organ physiology has been extensively discussed [77]. It was shown to have an indispensable role in the condition of specific metabolic needs, which explains the importance of peroxisomes in human congenital diseases. However, it becomes apparent that the amounts and properties of this organelle with respect to other organelles (e.g., ER, mitochondria) impact other pathologies of cell homeostasis, such as neurodegeneration, obesity, and more [2, 78].

Over the last decade, our knowledge of human diseases has drastically increased due to the breakthrough in sequencing technologies [79–81]. Projects such as the 1000 Genomes Initiative, ClinVar, OMIM, gnomAD, and others, provide the genetic variation landscape across individuals and populations [82–85]. Databases from such projects and large biobanks (e.g., UK BioBank [86]) are successfully being used for linking genetics with human diseases. For example, the UK BioBank resource gathered genotyping and exome sequencing data of approximately 500,000 people, combining it with clinical and lifestyle information. The unprecedented quality of these resources, merged with computational solutions, data sharing, and standardization advanced the field of human diseases. We briefly introduce computational-based methodologies used for improving the utility of genetic variations in the case of genetic-based peroxisomal diseases.

5.1 Advances in human genetics research

Mendelian diseases are caused by pathological mutations in a single gene with high penetrance. Consequently, the manifestation of the disease is determined by the simple rules of dominant or recessive inheritance. On the other hand, complex diseases result from the presence of many variants where each may carry a small effect. To study such diseases, genome-wide association studies (GWAS) have been used to connect human genetics with complex diseases. The ultimate goal of GWAS is to identify causal connections between genetic variants, traits, and phenotypes. GWAS provides a statistical value to a genotyped variant in the genome to assess the contribution of any specific variants to the studied phenotype. GWAS is useful to suggest association in cases where the sample size, the allele frequency, and the effect size of the association reach statistical significance [87]. Unfortunately, GWAS is prone to confounding factors and biases due to unresolved population structure and linkage disequilibrium (LD) [88, 89]. To overcome the drawback of GWAS due to population size, family-based studies are used as an attractive alternative. In such cases, the genetic variations of family members are determined (i.e., genotyping of parents and their inflicted child).

Another subfield in human genetics that is likely to impact modern medicine is polygenic risk scores (PRS). For PRS, genetics and data from health records are used to present a model for individuals' risk of having the disease of interest [90]. To make a meaningful prediction, the PRS model aggregates an individual's genotype information. Converting PRS to a machine learning prediction model calls for large-scale individual-level data, often using the summary statistics of GWAS results to build such a PRS model [91]. Currently, PRS is not yet typically incorporated into clinical settings and its utility remains questionable [92]. With the improvement in sequencing quality and amounts, rare variants turned out to carry special interest and importance [93, 94]. They occur at a low frequency, still exerting strong phenotypic effects. It is expected that the contribution of rare variants may be substantial in explaining disease heritability. The release of 500,000 whole-exome sequences (WES) by the UK-Biobank allows us to expand our knowledge including the genetic basis of relatively rare occurring diseases [95]. Burden tests in which genes, or defined chromosomal segments, replace individual variants as the statistical relevant unit improve genetic interpretability and utility [96]. For example, under the PWAS (proteome-based association study) methodology, coding genes are associated with a disease by quantifying the effect of genetic variation on the protein function. To assign a valid association between genetic results and the medical condition, PWAS requests that the impact of the genetic variations in the disease and healthy cohorts will be significantly different [97].

Peroxisomal disease does not always fit a trivial definition of a simple Mendelian disease, nor does it match chronic complex disease (e.g., type 2 diabetes). Peroxisomal diseases occur with a wide range of symptoms, interaction with developmental disorders, and severity. Therefore, studying the genetic basis of peroxisomal diseases covers two molecular etiologies—(i) The genetics of PEX proteins with a cellular understanding of organelle biology (e.g., peroxisome formation); (ii) Compilation of an exhaustive list of genetic variants and genes associated with peroxisomes. In this section, we only address the genetic approach associated with these two research goals.

5.2 Classes of peroxisomal diseases

Peroxisomes play a critical role in a variety of metabolic processes, especially in lipid metabolism. All active cells contain many peroxisomes (100 s to 1000s), therefore displaying varying capacity toward metabolic homeostasis, oxidative stress, and lipid metabolism in general. A failure in producing functional peroxisomes is the cause of numerous genetic diseases [98]. Conversely, one or more genetic defects too may result in failure of functional peroxisome production. Peroxisomal disorders are classified into two groups: (i) Specific peroxisomal enzyme deficiencies; (ii) Peroxisome biogenesis disorders (PBDs). PBDs result from a failure in the post-translational import of proteins to the peroxisome's matrix. The underlying genetics was used to better understand peroxisome formation while focusing on the failure to regulate multiple processes in peroxisomal cellular dynamics. In this section, we briefly categorized the peroxisomal diseases that are directly attributed to the failure to produce functional organelle. These diseases are connected to defects in the recognition of newly synthesized proteins, defects in the synthesis of peroxisome membranes, and failure in the insertion into the peroxisomal membrane (**Figure 3**).

PBDs display a spectrum of related diseases that differ in their severity, clinical manifestations, and underlying genetics. PBDs with clear molecular causal genes are collectively called Zellweger spectrum syndromes (ZSS). This set includes Zellweger syndrome (ZS), infantile Refsum disease (IRD), and neonatal adrenoleukodystrophy (NALD). The class of peroxisomal diseases that are caused by enzymatic deficiencies includes, for example, rhizomelic chondrodysplasia punctata (RCDP) type 2. Altogether there are 12 known rare diseases whose protein deficiencies are known, simple biomarkers for definitive diagnosis standards are missing. The task of

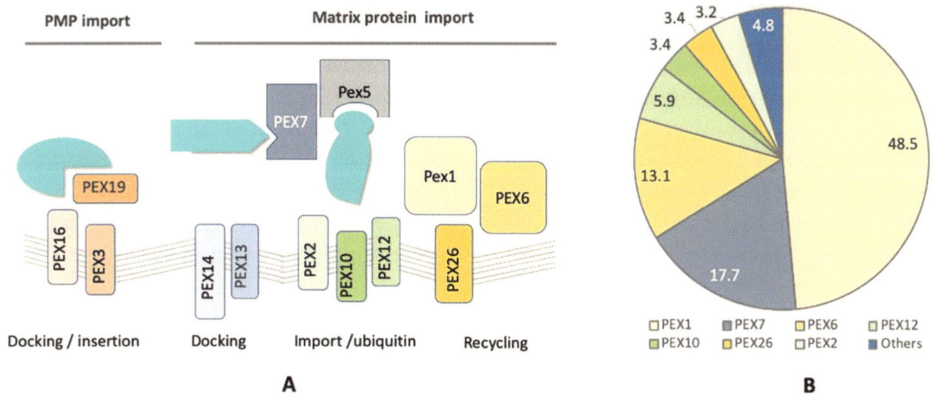

Figure 3.
Molecular basis of peroxisomal diseases. (a) Molecular view of the peroxisomal membrane with the main PEX proteins. PEX5 and PEX7 are receptors for the PTS1 and PTS2 proteins, respectively. PEX19 binds to proteins for insertion in the peroxisome membrane. Light blue indicates cargo proteins. (b) Frequency distribution of mutations in PEX proteins listed in a. the percentage is based on > 1300 patients diagnosed with peroxisome biogenesis disorder (PBDs). Adapted from [99]).

diagnosis of the different peroxisomal diseases was empowered by a machine learning algorithm [100].

In most ZSS diseases, very long-chain fatty acids and branched-chain fatty acids accumulate in the plasma of the affected individuals. The accumulation of these lipids impairs multiple organs, leading to a poor prognosis. Among the ZSS diseases, Zellweger syndrome (ZS) accounts for most cases. ZS is a rare autosomal recessive disorder (1:50,000) that in most cases is caused by mutations in PEX1, a member of the AAA-type ATPase family. Mechanistically, the disease is caused by defective protein import due to the mutations in one of 13 major PEX proteins (**Figure 3a**). The partition between patients with isolated deficiencies of metabolic enzymes and those with ZSS can be clinically challenging. Thus, a set of biochemical assays were developed to monitor peroxisome functions and facilitate correct diagnosis (e.g., levels of alanine transaminase or alkaline phosphatase). In PBDs, the reported mutations occur in any of the 13 major PEX proteins at different frequencies (**Figure 3b**). The interpretation of the mutations' effect on the peroxisome function is based on the accumulated knowledge of the role of PEX genes in the biogenesis of the organelle. Altogether, most peroxisomal human diseases are associated with a failure of the PEX proteins to import peroxisomal matrix proteins. In healthy individuals, newly synthesized matrix proteins reach the peroxisome by interacting with PEX5 or PEX7, cytosolic receptors that recognize either a C′-terminal PTS1 or an N′-terminal PTS2. In humans, two alternatively spliced PEX5 isoforms coexist, but only the longer version also binds PEX7. PEX2, PEX10, and PEX12 are zinc-binding RING finger proteins that are responsible for the ubiquitination of PEX5 which is essential for its recycling. The membranous peroxisomal membrane PEX26 combined with PEX1 and PEX6 is involved in the recycling process by releasing ubiquitinated PEX5 from the membrane. The malfunction of PEX7 results in a clinical phenotype of RCDP type 1 disease.

The identification of the genes underlying PBDs initially relied on detailed studies in yeast using classical genetics complementation groups [101]. With the availability of genotyping and WES data, the field of genetic testing evolved, and currently, searching for pathological mutations in PEX genes is used as a diagnostic service [99]. It is important to note that some of the genetic mutations are not restricted to PEX proteins but are also associated with genes involving the dynamics of peroxisomes and other organelles (e.g., mitochondrial and peroxisomal fission). Remarkably, cells with intact and functional proteins that do not reach the lumen of the peroxisome, result in severe phenotypes, presumably due to the rapid degradation of these misallocated enzymes. This is a general trend among many of the peroxisome matrix proteins.

Mutations in PEX1 are associated with the majority of ZSS cases. PEX1 contains ATP-binding motifs with ATPase activity. All the mutations reported are either defined as loss of function, but others are single missense mutations that affect the ATP binding pocket of the protein interface of PEX1 and PEX6. For a detailed genetic summary of the other PEX proteins, see ref. [99]. In contrast, X-linked adrenoleukodystrophy (ALD), an X-linked disorder, is caused by mutations in the ABCD1 gene that encodes an ABC transporter that act as a channel for the very-long-chain fatty acids entering the peroxisome.

5.3 Genomics tools for the study of peroxisome biology

A valuable resource called OpenTargets integrates multiple large-scale omics data including literature, drugs, and pathways [102]. The goal of OpenTargets is to bridge between molecular targets, drugs, and human diseases. Under the term of peroxisomal diseases in OpenTargets, many of the single enzyme defects that specify alteration in the metabolic enzymes are included. The strength of such a resource is in the ability in exposed overlooked processes that are dependent on intact peroxisome function. For example, among the 35 significant gene targets, all major PEX proteins are included (**Figure 3**). However, the strong support for human diseases that involve PEX11B, indicates that peroxisomal fission is a key process for cell homeostasis as PEX11B acts in recruiting dynamin-related GTPase to the peroxisomal membrane [103].

Based on GWAS results and support from mouse models the list of candidate genes for peroxisomal diseases keeps expanding. Currently, most of the peroxisomal mutated metabolic enzymes are listed, for example, acyl-CoA oxidase 1 (ACOX1), alanine-glyoxylate, serine-pyruvate aminotransferase (AGXT), and many more. Importantly, through an integrative approach that combines medical case studies and model organisms, several candidate genes whose function in peroxisome biology was not established are scored high, suggesting their overlooked roles in peroxisomal function, for example, UniProtKB entries E9PAM4 (Phosphatidylinositol 4-Kinase Type 2) and E9PPB4 (Peroxisomal Biogenesis Factor 19 Isoform 2). In summary, the list of variants associated with peroxisomal diseases was instrumental in shedding light on peroxisome metabolism and dynamics.

6. Conclusions

The last two decades have seen a fast-growing interest in the understanding of peroxisome biology with works on peroxisome biogenesis, import mechanisms, and the characterization of peroxisome proteins. It also emphasizes the existing gap between the more advanced methodologies of protein research concerning the understudied field of lipidomics. In parallel, the power of computational approaches will become pivotal in answering still-open questions regarding regulation, metabolism, and inter-organelle communication at a high level, and the elucidation of the mechanism and structure of peroxisome proteins and complexes at a more detailed level. We have aimed to provide an overview of some of the most important computational approaches that are likely to serve the research community, both basic and clinical, to expand its research toolbox in the study of peroxisome biology in health and disease.

Acknowledgements

The study was supported by the Israel Science Foundation (ISF) grant (2753/20) and CIRD (3035000323). We thank Dan Ofer for critical reading and Prof. Maya Schuldiner and Dr. Einat Zalckvar (Weizmann Institute) for sharing their insights on peroxisome biology.

Conflict of interest

The authors declare no conflict of interest.

Abbreviations

ER	Endoplasmic reticulum
GWAS	Genome-wide association study
MFA	Metabolic flux analysis
MS	mass spectrometry
PEX	Peroxin
PMP	Peroxisomal membrane protein
PTS	Peroxisomal targeting signal
PPI	protein–protein interactions
PTM	post-translational modification
WES	Whole-exome sequencing
ZSS	Zellweger spectrum syndromes

Author details

Naomi van Wijk and Michal Linial*
Department of Biological Chemistry, The Alexander Silberman Institute of Life Sciences, The Hebrew University of Jerusalem, Jerusalem, Israel

*Address all correspondence to: michall@cc.huji.ac.il

References

[1] Islinger M, Grille S, Fahimi HD, Schrader M. The peroxisome: An update on mysteries. Histochemistry and Cell Biology. 2012;**137**(5):547-574

[2] Islinger M, Voelkl A, Fahimi HD, Schrader M. The peroxisome: An update on mysteries 2.0. Histochemistry and Cell Biology. 2018;**150**(5):443-471

[3] Lu M, Ward E, van Tartwijk FW, Kaminski CF. Advances in the study of organelle interactions and their role in neurodegenerative diseases enabled by super-resolution microscopy. Neurobiology of Disease. 2021;**159**:105475

[4] Moen E, Bannon D, Kudo T, Graf W, Covert M, Van Valen D. Deep learning for cellular image analysis. Nature Methods. 2019;**16**(12):1233-1246

[5] Thul PJ, Åkesson L, Wiking M, Mahdessian D, Geladaki A, Ait Blal H, et al. A subcellular map of the human proteome. Science. 2017;**356**(6340):eaal3321

[6] Fransen M, Lismont C, Walton P. The peroxisome-mitochondria connection: How and why? IJMS. 2017;**18**(6):1126

[7] Shai N, Schuldiner M, Zalckvar E. No peroxisome is an island—Peroxisome contact sites. Biochimica et Biophysica Acta (BBA): Molecular Cell Research. 2016;**1863**(5):1061-1069

[8] Nixon-Abell J, Obara CJ, Weigel AV, Li D, Legant WR, Xu CS, et al. Increased spatiotemporal resolution reveals highly dynamic dense tubular matrices in the peripheral ER. Science. 2016;**354**(6311):aaf3928

[9] Li J, Peng J, Jiang X, Rea AC, Peng J, Hu J. DeepLearnMOR: A deep-learning framework for fluorescence image-based classification of organelle morphology. Plant Physiology. 2021;**186**(4):1786-1799

[10] Godinez WJ, Hossain I, Lazic SE, Davies JW, Zhang X. A multi-scale convolutional neural network for phenotyping high-content cellular images. Bioinformatics. 2017;**33**(13):2010-2019

[11] Xu Y-Y, Yang F, Shen H-B. Incorporating organelle correlations into semi-supervised learning for protein subcellular localization prediction. Bioinformatics. 2016;**32**(14):2184-2192

[12] Morone D, Marazza A, Bergmann TJ, Molinari M. Deep learning approach for quantification of organelles and misfolded polypeptide delivery within degradative compartments. MBoC. 2020;**31**(14):1512-1524

[13] Falk T, Mai D, Bensch R, Çiçek Ö, Abdulkadir A, Marrakchi Y, et al. U-Net: Deep learning for cell counting, detection, and morphometry. Nature Methods. 2019;**16**(1):67-70

[14] Yudistira N, Kavitha M, Itabashi T, Iwane AH, Kurita T. Prediction of sequential organelles localization under imbalance using a balanced deep U-Net. Scientific Reports. 2020;**10**(1):2626

[15] Jaqaman K, Loerke D, Mettlen M, Kuwata H, Grinstein S, Schmid SL, et al. Robust single-particle tracking in live-cell time-lapse sequences. Nature Methods. 2008;**5**(8):695-702

[16] Tinevez J-Y, Perry N, Schindelin J, Hoopes GM, Reynolds GD, Laplantine E, et al. TrackMate: An open and extensible platform for single-particle tracking. Methods. 2017;**115**:80-90

[17] Valm AM, Cohen S, Legant WR, Melunis J, Hershberg U, Wait E, et al. Applying systems-level spectral imaging and analysis to reveal the organelle interactome. Nature. 2017;**546**(7656):162-167

[18] Ounkomol C, Seshamani S, Maleckar MM, Collman F, Johnson GR. Label-free prediction of three-dimensional fluorescence images from transmitted-light microscopy. Nature Methods. 2018;**15**(11):917-920

[19] Brent R, Boucheron L. Deep learning to predict microscope images. Nature Methods. 2018;**15**(11):868-870

[20] Callaway E. 'It will change everything': DeepMind's AI makes gigantic leap in solving protein structures. Nature. 2020;**588**(7837):203-204

[21] Jumper J, Evans R, Pritzel A, Green T, Figurnov M, Ronneberger O, et al. Highly accurate protein structure prediction with AlphaFold. Nature. 2021;**596**(7873):583-589

[22] Varadi M, Anyango S, Deshpande M, Nair S, Natassia C, Yordanova G, et al. AlphaFold protein structure database: Massively expanding the structural coverage of protein-sequence space with high-accuracy models. Nucleic Acids Research. 2022;**50**(D1):D439-D444

[23] Mariani V, Biasini M, Barbato A, Schwede T. lDDT: A local superposition-free score for comparing protein structures and models using distance difference tests. Bioinformatics. 2013;**29**(21):2722-2728

[24] Jansen RLM, Santana-Molina C, van den Noort M, Devos DP, van der Klei IJ. Comparative genomics of peroxisome biogenesis proteins: Making sense of the PEX proteins. Frontiers in Cell and Development Biology. 2021;**20**(9):654163

[25] Ruff KM, Pappu RV. AlphaFold and implications for intrinsically disordered proteins. Journal of Molecular Biology. 2021;**433**(20):167208

[26] Pan D, Nakatsu T, Kato H. Crystal structure of peroxisomal targeting signal-2 bound to its receptor complex Pex7p–Pex21p. Nature Structural & Molecular Biology. 2013;**20**(8):987-993

[27] Weitzner BD, Jeliazkov JR, Lyskov S, Marze N, Kuroda D, Frick R, et al. Modeling and docking of antibody structures with Rosetta. Nature Protocols. 2017;**12**(2):401-416

[28] Bonneau R, Tsai J, Ruczinski I, Chivian D, Rohl C, Strauss CE, et al. Rosetta in CASP4: Progress in ab initio protein structure prediction. Proteins. 2001;**5**:119-126

[29] Bonneau R, Strauss CEM, Rohl CA, Chivian D, Bradley P, Malmström L, et al. De novo prediction of three-dimensional structures for major protein families. Journal of Molecular Biology. 2002;**322**(1):65-78

[30] Bradley P, Misura KMS, Baker D. Toward high-resolution de novo structure prediction for small proteins. Science. 2005;**309**(5742):1868-1871

[31] Yarov-Yarovoy V, Schonbrun J, Baker D. Multipass membrane protein structure prediction using Rosetta. Proteins. 2005;**62**(4):1010-1025

[32] Barth P, Schonbrun J, Baker D. Toward high-resolution prediction and design of transmembrane helical protein structures. Proceedings of the National Academy of Sciences of the United States of America. 2007;**104**(40):15682-15687

[33] Barth P, Wallner B, Baker D. Prediction of membrane protein structures with complex topologies

using limited constraints. Proceedings of the National Academy of Sciences of the United States of America. 2009;**106**(5):1409-1414

[34] Baek M, DiMaio F, Anishchenko I, Dauparas J, Ovchinnikov S, Lee GR, et al. Accurate prediction of protein structures and interactions using a three-track neural network. Science. 2021;**373**(6557):871-876

[35] Pennisi E. Protein structure prediction now easier, faster. Science. 2021;**373**(6552):262-263

[36] Yang J, Anishchenko I, Park H, Peng Z, Ovchinnikov S, Baker D. Improved protein structure prediction using predicted interresidue orientations. Proceedings of the National Academy of Sciences of the United States of America. 2020;**117**(3):1496-1503

[37] Schneidman-Duhovny D, Wolfson HJ. Modeling of Multimolecular Complexes. Methods in Molecular Biology. 2020;**2112**:163-174

[38] Rosell M, Fernández-Recio J. Docking approaches for modeling multi-molecular assemblies. Current Opinion in Structural Biology. 2020;**64**:59-65

[39] Evans R, O'Neill M, Pritzel A, Antropova N, Senior A, Green T, et al. Protein complex prediction with AlphaFold-Multimer [Internet]. Bioinformatics. 2021 [cited 2022 Jan 27]. Available from: http://biorxiv.org/lookup/doi/10.1101/2021.10.04.463034

[40] Tsaban T, Varga JK, Avraham O, Ben-Aharon Z, Khramushin A, Schueler-Furman O. Harnessing protein folding neural networks for peptide–protein docking. Nature Communications. 2022;**13**(1):176

[41] Humphreys IR, Pei J, Baek M, Krishnakumar A, Anishchenko I, Ovchinnikov S, et al. Computed structures of core eukaryotic protein complexes. Science. 2021;**374**(6573):eabm4805

[42] Masrati G, Landau M, Ben-Tal N, Lupas A, Kosloff M, Kosinski J. Integrative structural biology in the era of accurate structure prediction. Journal of Molecular Biology. 2021;**433**(20):167127

[43] Flower TG, Hurley JH. Crystallographic molecular replacement using an in silico-generated search model of SARS-CoV-2 ORF8. Protein Science. 2021;**30**(4):728-734

[44] Callaway E. Revolutionary cryo-EM is taking over structural biology. Nature. 2020;**578**(7794):201-201

[45] Malhotra S, Träger S, Dal Peraro M, Topf M. Modelling structures in cryo-EM maps. Current Opinion in Structural Biology. 2019;**58**:105-114

[46] Kim SJ, Fernandez-Martinez J, Nudelman I, Shi Y, Zhang W, Raveh B, et al. Integrative structure and functional anatomy of a nuclear pore complex. Nature. 2018;**555**(7697):475-482

[47] Punjani A, Rubinstein JL, Fleet DJ, Brubaker MA. cryoSPARC: Algorithms for rapid unsupervised cryo-EM structure determination. Nature Methods. 2017;**14**(3):290-296

[48] ElGamacy M, Riss M, Zhu H, Truffault V, Coles M. Mapping local conformational landscapes of proteins in solution. Structure. 2019;**27**(5):853-865.e5

[49] Piersimoni L, Kastritis PL, Arlt C, Sinz A. Cross-linking mass spectrometry for investigating protein conformations and protein–protein interactions: A method for all seasons. Chemical Review. 2021;acs.chemrev.1c00786.

[50] Slavin M, Zamel J, Zohar K, Eliyahu T, Braitbard M, Brielle E, et al. Targeted in situ cross-linking mass spectrometry and integrative modeling reveal the architectures of three proteins from SARS-CoV-2. Proceedings of the National Academy of Sciences of the United States of America. 2021;**118**(34):e2103554118

[51] Yates JR III, Gilchrist A, Howell KE, Bergeron JJM. Proteomics of organelles and large cellular structures. Nature Reviews. Molecular Cell Biology. 2005;**6**(9):702-714

[52] Drissi R, Dubois M-L, Boisvert F-M. Proteomics methods for subcellular proteome analysis. The FEBS Journal. 2013;**280**(22):5626-5634

[53] Yifrach E, Fischer S, Oeljeklaus S, Schuldiner M, Zalckvar E, Warscheid B. Defining the mammalian peroxisomal proteome. In: del Río LA, Schrader M, editors. Proteomics of Peroxisomes [Internet]. Singapore: Springer; 2018. p. 47-66

[54] Neilson KA, Ali NA, Muralidharan S, Mirzaei M, Mariani M, Assadourian G, et al. Less label, more free: Approaches in label-free quantitative mass spectrometry. Proteomics. 2011;**11**(4):535-553

[55] Collins BC, Hunter CL, Liu Y, Schilling B, Rosenberger G, Bader SL, et al. Multi-laboratory assessment of reproducibility, qualitative and quantitative performance of SWATH-mass spectrometry. Nature Communications. 2017;**8**(1):291

[56] Mallick P, Schirle M, Chen SS, Flory MR, Lee H, Martin D, et al. Computational prediction of proteotypic peptides for quantitative proteomics. Nature Biotechnology. 2007;**25**(1): 125-131

[57] Gillet LC, Navarro P, Tate S, Röst H, Selevsek N, Reiter L, et al. Targeted data extraction of the MS/MS spectra generated by data-independent acquisition: A new concept for consistent and accurate proteome analysis. Molecular & Cellular Proteomics. 2012;**11**(6)

[58] Doerr A. Mass spectrometry–based targeted proteomics. Nature Methods. 2013;**10**(1):23-23

[59] Basic Sciences Division, Universidad de Monterrey, San Pedro Garza García, N.L. Mexico, Gracia KC, Husi H, Institute of Cardiovascular and Medical Sciences, BHF Glasgow Cardiovascular Research Centre, University of Glasgow, Glasgow, UK, Division of Biomedical Sciences, Centre for Health Science, University of Highlands and Islands, Inverness, UK. Computational Approaches in Proteomics. In: Division of Biomedical Science, University of the Highlands and Islands, UK, Husi H, editors. Computational Biology [Internet]. Codon Publications; 2019 [cited 2022 Jan 30]. p. 119-42. Available from: https://exonpublications.com/index.php/exon/article/view/223

[60] Krokhin OV, Craig R, Spicer V, Ens W, Standing KG, Beavis RC, et al. An improved model for prediction of retention times of tryptic peptides in ion pair reversed-phase HPLC. Molecular & Cellular Proteomics. 2004;**3**(9):908-919

[61] Krokhin OV, Ying S, Cortens JP, Ghosh D, Spicer V, Ens W, et al. Use of peptide retention time prediction for protein identification by off-line reversed-phase HPLC–MALDI MS/MS. Analytical Chemistry. 2006;**78**(17):6265-6269

[62] Wen B, Zeng W, Liao Y, Shi Z, Savage SR, Jiang W, et al. Deep Learning in Proteomics. Proteomics. Vol. 202020. p. 1900335

[63] Wiese S, Gronemeyer T, Ofman R, Kunze M, Grou CP, Almeida JA, et al. Proteomics characterization of mouse kidney peroxisomes by Tandem mass spectrometry and protein correlation profiling. Molecular & Cellular Proteomics. 2007;**6**(12):2045-2057

[64] Andersen JS, Mann M. Organellar proteomics: Turning inventories into insights. EMBO Reports. 2006;**7**(9):874-879

[65] Itzhak DN, Tyanova S, Cox J, Borner GH. Global, quantitative and dynamic mapping of protein subcellular localization. eLife. 2016;**9**(5):e16950

[66] Kunze M. Predicting peroxisomal targeting signals to elucidate the peroxisomal proteome of mammals. In: del Río LA, Schrader M, editors. Proteomics of Peroxisomes [Internet]. Singapore: Springer. p. 157-199

[67] Neuberger G, Maurer-Stroh S, Eisenhaber B, Hartig A, Eisenhaber F. Prediction of peroxisomal targeting signal 1 containing proteins from amino acid sequence. Journal of Molecular Biology. 2003;**328**(3):581-592

[68] Neuberger G, Maurer-Stroh S, Eisenhaber B, Hartig A, Eisenhaber F. Motif refinement of the peroxisomal targeting signal 1 and evaluation of taxon-specific differences. Journal of Molecular Biology. 2003;**328**(3): 567-579

[69] Luck K, Sheynkman GM, Zhang I, Vidal M. Proteome-scale human interactomics. Trends in Biochemical Sciences. 2017;**42**(5):342-354

[70] Qin Y, Huttlin EL, Winsnes CF, Gosztyla ML, Wacheul L, Kelly MR, et al. A multi-scale map of cell structure fusing protein images and interactions. Nature. 2021;**600**(7889):536-542

[71] Čuperlović-Culf M, Barnett DA, Culf AS, Chute I. Cell culture metabolomics: Applications and future directions. Drug Discovery Today. 2010;**15**(15-16):610-621

[72] Schrimpe-Rutledge AC, Codreanu SG, Sherrod SD, McLean JA. Untargeted metabolomics strategies—Challenges and emerging directions. Journal of the American Society for Mass Spectrometry. 2016;**27**(12):1897-1905

[73] Orth JD, Thiele I, Palsson BØ. What is flux balance analysis? Nature Biotechnology. 2010;**28**(3):245-248

[74] Lagziel S, Lee WD, Shlomi T. Studying metabolic flux adaptations in cancer through integrated experimental-computational approaches. BMC Biology. 2019;**17**(1):51

[75] Lee WD, Mukha D, Aizenshtein E, Shlomi T. Spatial-fluxomics provides a subcellular-compartmentalized view of reductive glutamine metabolism in cancer cells. Nature Communications. 2019;**10**(1):1351

[76] Becker SA, Feist AM, Mo ML, Hannum G, Palsson BØ, Herrgard MJ. Quantitative prediction of cellular metabolism with constraint-based models: The COBRA Toolbox. Nature Protocols. 2007;**2**(3):727-738

[77] Schrader M, Costello J, Godinho LF, Islinger M. Peroxisome-mitochondria interplay and disease. Journal of Inherited Metabolic Disease. 2015;**38**(4):681-702

[78] Argyriou C, D'Agostino MD, Braverman N. Peroxisome biogenesis disorders. TRD. 2016;**1**(2):111-144

[79] Lappalainen T, Scott AJ, Brandt M, Hall IM. Genomic analysis in the age of human genome sequencing. Cell. 2019;**177**(1):70-84

[80] van Dijk EL, Jaszczyszyn Y, Naquin D, Thermes C. The third revolution in sequencing technology. Trends in Genetics. 2018;**34**(9):666-681

[81] Warr A, Robert C, Hume D, Archibald A, Deeb N, Watson M. Exome sequencing: Current and future perspectives. G3 Genes|Genomes|Genetics. 2015;**5**(8):1543-1550

[82] Auton A, Abecasis GR, Steering Committee, Altshuler DM, et al. A global reference for human genetic variation. Nature. 2015;**526**(7571):68-74

[83] Landrum MJ, Lee JM, Benson M, Brown GR, Chao C, Chitipiralla S, et al. ClinVar: Improving access to variant interpretations and supporting evidence. Nucleic Acids Research. 2018;**46**(D1):D1062-D1067

[84] Amberger JS, Bocchini CA, Schiettecatte F, Scott AF, Hamosh A. OMIM.org: Online Mendelian Inheritance in Man (OMIM®), an online catalog of human genes and genetic disorders. Nucleic Acids Research. 2015;**43**:D789-D798

[85] Karczewski KJ, Francioli LC, Tiao G, Cummings BB, Alföldi J, Wang Q, et al. The mutational constraint spectrum quantified from variation in 141,456 humans. Nature. 2020;**581**(7809):434-443

[86] Bycroft C, Freeman C, Petkova D, Band G, Elliott LT, Sharp K, et al. The UK Biobank resource with deep phenotyping and genomic data. Nature. 2018;**562**(7726):203-209

[87] Bush WS, Moore JH. Chapter 11: Genome-wide association studies. PLoS Computational Biology. 2012;**8**(12):e1002822

[88] Lee JJ, McGue M, Iacono WG, Chow CC. The accuracy of LD Score regression as an estimator of confounding and genetic correlations in genome-wide association studies. Genetic Epidemiology. 2018;**42**(8):783-795

[89] Sul JH, Martin LS, Eskin E. Population structure in genetic studies: Confounding factors and mixed models. Barsh GS, editor. PLoS Genetics. 2018;**14**(12):e1007309

[90] Duncan L, Shen H, Gelaye B, Meijsen J, Ressler K, Feldman M, et al. Analysis of polygenic risk score usage and performance in diverse human populations. Nature Communications. 2019;**10**(1):3328

[91] Wand H, Lambert SA, Tamburro C, Iacocca MA, O'Sullivan JW, Sillari C, et al. Improving reporting standards for polygenic scores in risk prediction studies. Nature. 2021;**591**(7849):211-219

[92] Torkamani A, Wineinger NE, Topol EJ. The personal and clinical utility of polygenic risk scores. Nature Reviews: Genetics. 2018;**19**(9):581-590

[93] Guo MH, Plummer L, Chan Y-M, Hirschhorn JN, Lippincott MF. Burden testing of rare variants identified through exome sequencing via publicly available control data. The American Journal of Human Genetics. 2018;**103**(4):522-534

[94] Cirulli ET, White S, Read RW, Elhanan G, Metcalf WJ, Tanudjaja F, et al. Genome-wide rare variant analysis for thousands of phenotypes in over 70,000 exomes from two cohorts. Nature Communications. 2020;**11**(1):542

[95] Nait Saada J, Kalantzis G, Shyr D, Cooper F, Robinson M, Gusev A, et al. Identity-by-descent detection across 487,409 British samples reveals fine scale population structure and ultra-rare variant associations. Nature Communications. 2020;**11**(1):6130

[96] Liu DJ, Peloso GM, Zhan X, Holmen OL, Zawistowski M, Feng S, et al. Meta-analysis of gene-level tests for rare variant association. Nature Genetics. 2014;**46**(2):200-204

[97] Brandes N, Linial N, Linial M. PWAS: Proteome-wide association study—linking genes and phenotypes by functional variation in proteins. Genome Biology. 2020;**21**(1):173

[98] Braverman NE, D'Agostino MD, MacLean GE. Peroxisome biogenesis disorders: Biological, clinical and pathophysiological perspectives: Peroxisome biogenesis disorders. Developmental Disabilities Research Reviews. 2013;**17**(3):187-196

[99] Waterham HR, Ebberink MS. Genetics and molecular basis of human peroxisome biogenesis disorders. Biochimica et Biophysica Acta (BBA): Molecular Basis of Disease. 2012;**1822**(9):1430-1441

[100] Subhashini P, Jaya Krishna S, Usha Rani G, Sushma Chander N, Maheshwar Reddy G, Naushad SM. Application of machine learning algorithms for the differential diagnosis of peroxisomal disorders. The Journal of Biochemistry. 2019;**165**(1):67-73

[101] Fujiki Y, Okumoto K, Mukai S, Honsho M, Tamura S. Peroxisome biogenesis in mammalian cells. Frontiers in Physiology [Internet]. 2014;**5**. [cited 2022 Feb 6]. Available from: http://journal.frontiersin.org/article/10.3389/fphys.2014.00307/abstract

[102] Ghoussaini M, Mountjoy E, Carmona M, Peat G, Schmidt EM, Hercules A, et al. Open targets genetics: Systematic identification of trait-associated genes using large-scale genetics and functional genomics. Nucleic Acids Research. 2021;**49**(D1):D1311-D1320

[103] Li X, Gould SJ. The dynamin-like GTPase DLP1 is essential for peroxisome division and is recruited to peroxisomes in part by PEX11. The Journal of Biological Chemistry. 2003;**278**(19):17012-17020